Welcome to the Book of Making

The future is here, and it's digital. It's in the electronics that make devices work; the sensors, the motors, and the logic that controls them. It's in the 3D printed plastic parts, and the CNC-machined wood and aluminium. If you want to build something in the modern age, you need to be thinking digitally.

We've rounded up a few of our favourite makes previously published in HackSpace magazine (which is now part of Raspberry Pi Official Magazine), which should keep you entertained for an afternoon, a weekend, or longer. Fire up your 3D printer, soldering iron, or Raspberry Pi, and start making!

ANDREW GREGORY

Editor 🔘 andrew.gregory@raspberrypi.com

PROJECTS FOR MAKERS & HACKERS

BUILD A CUSTOM KEYBOARD WITH RASPBERRY PI PICO

CUSTOM CONTROLS WITH A 3D PRINTED BIG BUTTON

BOOK OF MAKING 2026

CONTROL PROGRAMMABLE LEDS FOR DAZZLING LIGHTING

EVERYTHING YOU NEED TO KNOW ABOUT BATTERY POWER

FROM THE MAKERS OF RASPBERRY PI OFFICIAL MAGAZINE

E
Andrew Gregory
🔘 andrew.gregory@raspberrypi.com

Sub Editors
David Higgs, Nicola King, Phil King

Advertising
Charlotte Milligan
🔘 charlotte.milligan@raspberrypi.com
📞 +44 (0)7725 368887

DESIGN

Critical Media
🔗 criticalmedia.co.uk

Head of Design
Lee Allen

Designers
Sam Ribbits, Olivia Mitchell

Raspberry Pi

Head of Design
Jack Willis

Designers
Sara Parodi, Natalie Turner

Illustrator
Sam Alder

Brand Manager
Brian O Halloran

CONTRIBUTORS

Ben Everard, Jo Hinchliffe, Rob Miles, Nicola King, Andrew Lewis, Aaed Musa, Phil King, Turi Scandurra, Tim Ritson, Karl Mose, Stéphane Godin, Dr Ross Hamilton

PUBLISHING

Publishing Director
Brian Jepson
🔘 brian.jepson@raspberrypi.com

Director of Communications
Helen Lynn

CEO
Eben Upton

DISTRIBUTION

Seymour Distribution Ltd
2 East Poultry Ave,
London EC1A 9PT
📞 +44 (0)207 429 4000

SUBSCRIPTIONS

Unit 6 The Enterprise Centre
Kelvin Lane, Manor Royal,
Crawley, West Sussex, RH10 9PE
📞 +44 (0)1293 312193
🔗 rpimag.co/subscribe
🔘 rpipress@subscriptionhelpline.co.uk

ISBN: 9781916868519
UPC: 725274044351 55

Contents

3D printing by hand

12

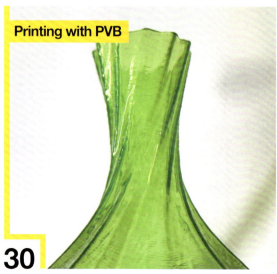
Printing with PVB

30

106

Solar tracker
130

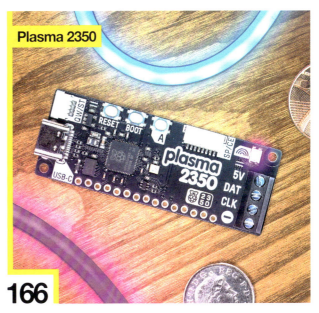
Plasma 2350
166

HackSpace

TECHNOLOGY IN YOUR HANDS

AFTERNOON PROJECTS

HACK | MAKE | BUILD | CREATE

When you have a couple of hours to yourself, these projects are quick to make – have fun!

PG 8

3D PRINTING

Make the most of your growing space with 3D printed planters

PG 12

3D PEN

Make squiggly shapes with melted plastic – it's 3D printing, but freehand

PG 16

LED LAMP

Tell the temperature with an RGB LED, a sensor and a Raspberry Pi Pico

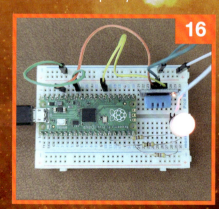

20

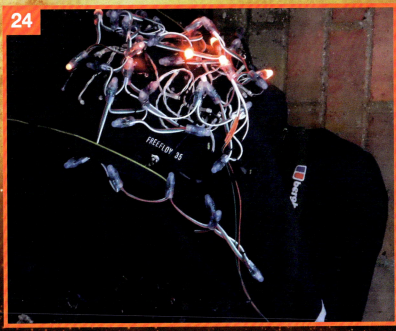

24

3D-printed planters

Make the most of your growing space

Ben Everard

Ben enjoys gardening because, unlike most of his projects, plants will complete the building process themselves (as long as he remembers to water them).

Right ◆
You can mix and match pots of different styles, sizes, and colours

Despite the best efforts of the gods of rain, the winter is finally coming to an end. Plants are sprouting all around – ones we planted and ones we haven't. This author has decided to turn his attention to generally making his home a little greener and much more flavourful.

Plants, like all life, are amazing replicators. Inside each seed is a tiny spool of cellulose filament and a solar-powered print head capable of squirting out this building block into exactly the right shape for that species (OK, we're computer nerds rather than biology nerds, and might have got that bit wrong). This makes us think that they're perfect partners for a bit of 3D printing.

The two big problems we've had with herbs are that there are quite a lot of them and they take up a lot of space (whether you want them to or not). One little mint plant very quickly becomes a whole carpet that, while delicious, covers half the garden.

There is a solution to both these problems – vertical gardening. This is the process of building your plants upwards rather than outwards. With this,

we can fit a lot of different plants into a small space and keep each contained in its own pot.

ONWARDS AND UPWARDS

A lot of vertical gardening is built around hydroponics, and if that's a route you want to go down, you'll find

Above ⬥
The PrusaVert
system uses old
filament spools
to add strength
to the planter

plenty of inspiration online. However, we wanted a more traditional soil-based approach for now. After perusing some repositories of 3D-printable models, we found a couple to try out.

PrusaVert (**hsmag.cc/ prusavert**) incorporates Prusament spools (which we just so happen to have a few spares of at the moment) to make a robust barrel-like vertical planter suitable for life outside, while High Gardens (**hsmag.cc/ highgardens**) makes it possible to tower-up plant pots on the windowsill.

Let's take a closer look at PrusaVert first. Prusament spools are two sides of plastic joined by cardboard in the middle. You can pull the two sides apart to remove the middle. You can buy replacement filament with just the cardboard, then slot the sides from the previous spools on. Alternatively, you can use these as the basis for

> **With this, we can fit a lot of different plants into a small** space and keep each contained in its own pot

additional builds, such as this vertical garden. Each tier consists of three 3D-printed plant pots that attach to one of the spool halves. They are designed to screw together, but we decided to try a drop of glue instead and, so far, they seem to be holding. Each tier can be joined together, but we have left them loose so they can be rearranged as needed. We think this will be useful when it comes to bringing some of them indoors to overwinter.

The pots come in two sizes – we went with the larger 12 cm-high version as these will hold more water and allow for bigger plants. Our plants seem happy and are quickly adapting to their new home.

CHOOSE YOUR OWN

High Gardens is composed of multiple pieces. You start with a base and then add as many pots as you like (or dare). Each pot slots into the one below it. →

Above ◆
The chicken run is in the sunniest spot in this garden, and now we can put that sun to good use

Right ◆
A hook on the planter goes over the plank to secure the pot in position

These slots let the water flow through the central pillar so, in theory at least, you can add water to a top reservoir and it will water each pot, with any excess draining into the tray at the bottom. We found that it didn't work particularly well as it was hard to gauge how to add water to the top so that all the pots would be watered without overflowing the tray at the bottom. We instead watered each pot individually, but we did find the water flow system useful as it meant that each plant could drain and the excess water would be captured.

We found that a tower roughly three or four pots high was stable on its own, but any higher seemed a bit precarious. This is partly due to the presence of two energetic kittens in the test area. We once ended up with plants scattered about the floor, but this was due to a kitten jumping up and clinging onto the blind, pulling herself along until she was above the High Gardens, and then dropping down onto it. If your environment is more sedate, you may get away with a higher garden. There are wall clips available (**hsmag.cc/hgclip**) which make the prospect of compost-covered floors a bit less likely.

There is a selection of different designs and sizes for both pots and bases, though some of the bases look alarmingly narrow. Perhaps some people like

TYPES OF **FILAMENT**

PLA isn't UV (ultraviolet) stable. This means that, over time, the UV rays in sunshine will gradually break it down. How fast this happens depends mainly on how much UV they are exposed to. Most plants like to be positioned outside or on windowsills which are, unfortunately, places where there's quite a bit of UV light.

Depending on what plants you grow, you might find that the foliage protects the plastic from UV. This might be sufficient to ensure that your planter lasts long enough, or it might not.

You can protect PLA with a layer of paint. Alternatively, you can use a UV-stable filament such as ASA.

to live dangerously, or perhaps some places aren't patrolled by feline vandals. The choice, as they say, is yours.

HANGING UP

Both the previous options build up from the floor, but sometimes you want to attach your pots to a wall. That's exactly what this vertical gardening planter does: **hsmag.cc/vertplant**. The attachment is simplicity itself – there's a hook on the planter, and you just need to secure a plank of wood to the wall with enough of a gap to hook the planter over. You can have as many or as few as you like hanging from your planks.

Misan, the designer of this planter, has a few versions with different levels of drainage. The v4 has intensive draining, with just a mesh at the bottom. This may be OK for succulents and other plants that like dry conditions, but moisture-loving plants may struggle. The v3, with its less-intensive draining, will be better suited for most plants.

This model just about prints successfully without supports, but we got better results when

LABELS

As well as printing the right plant pots, you can print other accessories for your garden. The most popular are plant labels. These can be as decorative or as plain as you like. Printing in two colours (either with an automatic colour switcher, or by switching filament at a particular Z height) is a great way to make striking labels that add a bit of colour to your veg beds.

we incorporated a few. It doesn't need a full set, but using PrusaSlicer's automatic support painting feature, we added just enough to ensure the long bridges printed well without wasting lots of filament.

We've been growing herbs indoors and outdoors in these for a few weeks, and so far, we're delighted with the results. We've got plenty of flavourful greenery to add to meals and drinks, and they look and smell wonderful as well. As the weather gets warmer, we'll print more and more of these to keep up with the growing demand. ◻

Above ◹
Each planter protrudes enough for the plants to get plenty of light, but if you place it against a wall, you might struggle to get plants to grow in some sections

3D pens

Is a 3D pen just a hot glue gun on steroids?
Let's take a closer look

Nicola King

Nicola King is a freelance writer and sub-editor. She tries to produce beautiful things but, as you'll see, sometimes they're not quite as beautiful as she had hoped…all part of the learning process.

Above ◆
A little wonky, and certainly unlike any frame you'd pick up in Specsavers… which is probably just as well or they'd go out of business

Meandering around the supermarket on our weekly shop a couple of months ago, we chanced upon a craft-related gizmo in the 'middle' aisles, priced at the princely sum of £24.99. This author's husband, always one for a new gadget, couldn't resist popping it in the trolley, citing the fact that it would be a great subject for a little tutorial, so we exited the grocery store with our broccoli, cheese, sausages, bin bags, and…oh, a 3D pen set.

Now, it's worth trying to manage some expectations from the get-go on this topic – a 3D pen will never (ever) be able to produce anything close to the quality of output of a 3D printer, but they can be fun to use and are arguably a great way of introducing younger members of the family to the concept of 3D printing. We read one review of a 3D pen online by someone who had tried one out with her young daughter to pique her interest, before going that step further and investing much more

heavily in a 3D printer. 3D printers are, let's be honest, on the expensive side and 3D pens are much kinder on your wallet, while still being able to bring your imagination to 3D life. With that in mind, let's investigate what you can expect from such a device.

FREEHAND 3D PRINTING… SORT OF

You might initially think that a 3D pen is something of a gimmick, but they can be used as genuine artistic tools that can help you sculpt in 3D, converting plastic filament into melted material with which you can create objects. So, if you want to create a 3D model by hand, these are worth a try. Here, we are bypassing the software that a 3D printer uses, and controlling it ourselves, with no difficult tech to master. Importantly, before we move on, let's also remember that we are working with molten plastic – it's hot, so keep your digits, and other parts of you, well away from the heated tip. That's especially important to remember when you are pulling any residue off the tip when you have finished, or are changing colours.

Above ◈
The contents of the kit that we purchased, including an impressive selection of filament colours for the price – ten colours of 5 m each in length, 1.75 mm in diameter. It incorporated everything that we needed to get underway

Left ◈
The USB cable is a good length, and a plus-point of this device is that it turns itself off when you set it aside after two minutes – always a handy feature. Just press the On/Off button to heat it up again

WHAT TO LOOK FOR WHEN PERUSING THE 3D PEN MARKET

So, what are the key things that you need to consider before buying?

• Comfort of use – 3D printing projects are not quick, and usually take a significant amount of time to undertake, so how the pen feels in your hand is very important (ours weighs 61 grams). You need it to be lightweight too, as well as ergonomically pleasing. Some pens are larger than others, wider or longer (ours was around 18 cm long, with a tip diameter of 0.7 mm).

• Reliability – you need a steady, even strand of filament coming out of the nozzle, at a comfortable speed, which is partly down to the dexterity of the user, but is also down to the quality of the pen itself. You don't want your pen prone to filament clogs.

• Does the pen come with everything you need, such as clear instructions and some templates, or even some filament? These are very useful to have for a newbie.

• Cost – what's your budget? Many 3D pens on the market won't break the bank.

• Finally, as with anything, it really comes down to who you are buying the pen for. If you consider yourself a high-end user who plans on using it a lot for design-type projects, and want all the bells and whistles, then you will need to pay for those features. For example, pens with digital displays where you can select the exact temperature you require will come at a more premium price. If you are overseeing younger makers, then you need to be conscious of the age-range specified on the box, the size of the pen, and it goes without saying that you need to be using a non-toxic filament at a lower temperature.

So, first things first, we attached the USB cable, and selected the PLA switch on the side of the device (the type of filament provided). Once the On/Off switch was pressed, we waited for the pen to heat up. Interestingly, it was really quick – after about 30 seconds, the pen had reached its required melting temperature (according to the accompanying booklet, around 190°C), the indicator light turned green, and we were ready for the off. Next, we chose the colour of filament that we wanted and fed it into the filament entrance with the straight-cut end (apparently that's important). After a few seconds, melted filament was coming out of the tip and we were able to get creating.

The truth is, at the risk of shattering some illusions, the absolutely best way to start is to begin by creating shapes in 2D on the flat… yes, we know it's a 3D pen but we have to start somewhere, we are beginners, and we are not going to run before we can walk. Plus, that's the advice we've picked up from online, so we decided to follow it.

Once you've got some 2D shapes, you can then weld them together to create a 3D shape. →

Left ◆
A hot mess of plastic – take your time to practise and get the feel of the pen and the speed of extrusion before you start creating

Below ◆
Place your clear mat over your template before you begin and, to prevent heat marks, hold the pen around 0.5 cm away. It's best not to touch the plastic mat with the tip

QUICK TIP

Protect your work surface from plastic residue and use a mat of some kind. A hobbyist cutting mat will do the job, and makes it easier to move your work around and lift it off.

Trying to draw something in the air, so that you really feel like you are creating a 3D shape, can be very tricky

FILAMENT FAMILIARITY

There are a few options to ponder when it comes to which meltable plastic to select for your 3D pen. Ours could use all three types detailed below. So, let's take a closer look:

- PLA (Polylactic Acid). This is arguably a top choice for use with 3D pens, for a number of worthy reasons. Firstly, it is a natural thermoplastic polyester derived from renewable resources such as soybeans, corn starch, or sugar cane, so it is a bioplastic. It is technically biodegradable (under very specific conditions) and sustainable. Many people find it easy to use PLA with 3D pens due to the low melting temperature, and it does not have a strong smell or give off lots of fumes, something that many makers are keen on avoiding if they can, and great if you are overseeing children using a 3D pen. It's available in many colours, even some that glow in the dark. You can even purchase 'wood filament' for some pens, which is not actually wood, but wood fibres added to a PLA base for a wood effect when extruded. PLA is often used in food packaging and plastic cups, so this underlines how safe it is. It is also rigid and strong; however, poor heat resistance properties make it mainly a hobbyist material – as it is made of organic materials, it has a higher permeability than some other options. If you are looking for longevity in your creations, PLA may also fall a little short.

- ABS (Acrylonitrile Butadiene Styrene). Now this filament is considered more flexible and tougher than PLA, as well as more durable, and is therefore great for 'bridging', i.e. extruding your filament with your pen between two points, with no support from below. Projects are also likely to last longer than PLA creations. However, ABS is not as environmentally friendly, as it is a petroleum-based thermoplastic and non-biodegradable. Because of this, when it's heated ABS can produce an unpleasant smell and any fumes are toxic, so you really need to ensure your ventilation is top-notch. You can use it in your pen; you just need to take the correct precautions. Also, consider that this filament can warp and shrink somewhat when you are working with it.

- PETG (Polyethylene Terephthalate Glycol). Some pens will also take PETG filament, which is safe, strong, and resilient, but it's not biodegradable. That said, it can be recycled.

When choosing your filament, you will also need to think about the filament diameter (many 3D pens use 1.75 mm), the total amount of filament required, tolerance, and compatibility with your pen, as well, of course, as the colours you want to utilise based on the project you are embarking on. Finally, it's worth mentioning that some 3D pen makers want you to use their own proprietary filament, which could be seen as a limitation as it may cost you more – just check this out before you buy.

PLASTIC PRECISION

We wanted some control over the filament flow, as when we first tried to use it, the filament came out way too fast and we had no clue what we were doing. So, we opted to operate the feed manually, which involved keeping the filament feeder button pressed down while we worked. If we wanted to stop the filament feed, we just released the button. There is also a speed switch on this pen, and three options to make the filament flow faster or slower. When you first start out, it's advisable to select a very slow speed.

We chose to create a pair of 'spectacles' as our first foray and, while they are obviously the work of a complete amateur, they held together when we welded the arms to the main body. A butterfly swiftly followed and we were well on a roll.

A STEP FURTHER

Now, trying to draw something in the air, so that you really feel like you are creating a 3D shape, can be very tricky and takes practice. For example, you could start off by trying to draw a cube. Begin with a square base, then from each corner draw a line upwards into the air (easier said than done!). Level them off and then you need to practise 'bridging' or connecting two points in space so that you can complete your cube. This is hard if the two points are not stable, and if you're new to using a 3D pen, this may be a little frustrating, which is why it's much easier to start in 2D, welding pieces together and then progressing to more challenging designs.

Another idea is to write a word (joined-up letters so it holds together) and then lean it against a hard surface. Then, with your pen, draw some struts out from either end so it can stand up. Have a play and see where you end up.

Above ⬥
This author's husband had a go and produced this very appropriate offering... we did say we need some practice!

Left ⬥
Mix your colours to practise changing pen colour filaments. It helps you get a feel for the various buttons on the pen too

QUICK TIP

Never leave a 3D pen unattended. Use it on a heat-resistant surface and don't use it near flammable materials – if these pens can melt plastic, they can melt other things too!

QUICK TIP

Don't over-extrude! By that, we mean don't push out too much filament when you are working on a project. Also, practise your hand speed, and don't move in a jerky manner as you work or that will result in unsightly blobs.

QUICK TIP

If you have pieces/ bulges on your design that you want to remove, you can use the tip of the nozzle to easily melt away small blips and errors. You can also use scissors for this, but maybe avoid using your best scissors in case they blunt.

EXTRUDING EVALUATION

In conclusion, we enjoyed trying out our very inexpensive 3D pen, but the phrase 'you get what you pay for' definitely comes to mind, and we'd be really interested to know what the slightly more expensive pens on the market are capable of – from the £50 to £100 price mark. Even though this was an inexpensive option, we were still pretty happy with the results and can see that practice and some time investment would definitely improve the finished articles. Be mindful that you will definitely not be able to create an exquisitely detailed model of the Statue of Liberty the first time that you use a 3D pen. It takes time and patience to learn how to keep your hand steady, to manipulate the filament, change colours and so on, but it's definitely worth having a go to see what you can create.

There are some who disparagingly call these cheaper-end pens 'toys', but seriously, what toy has a tip that hot? We feel that description is a little unfair and misleading, and prefer to think of the 3D pen as a creative tool that takes a little practice to master with a steady hand, and that just happens to be a lot of fun. □

MENDING MISPRINTS

If you are a stalwart 3D printer user, and have completely discounted ever using a 3D pen in any capacity, do bear in mind a rather handy aspect of owning one – you can use it to smooth out any issues on 3D prints from a 3D printer. Basically, it is very useful for 'gluing' or bonding pieces together, or touching up anything that your machine has printed that did not turn out quite how you hoped it would, so don't underestimate its usefulness. It's an alternative to superglue and worth a try to see how it works for you. We suspect ABS filament would give a stronger adhesion than PLA, but have a go and see how effective it is.

Assuming your filaments in the two devices are the same (which obviously makes colour-matching very easy), you can repair cracks, gaps, and holes in 3D-printed items. For example, two broken parts can be welded back together, as long as you work with a slow extrusion rate to allow more precise work. Any excess when you've finished can be trimmed off with scissors, and then you can sand it down.

Here's a handy link to give more insight: **hsmag.cc/3DPenWelding.**

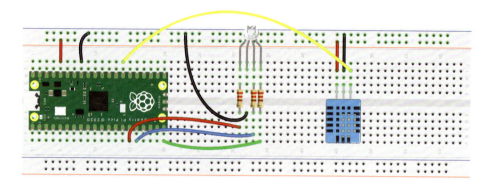

Temperature-controlled colour lamp

This easy-to-make electronics project combines Raspberry Pi Pico, temperature sensor, and RGB LED

Phil King

A long-time Raspberry Pi user and tinkerer, Phil is a freelance writer and editor with a focus on technology.

Figure 1 ◈
The wiring diagram for connecting an RGB LED and DHT11 sensor to the Raspberry Pi Pico; double-check the pin ordering on your LED and sensor, as it may differ

nce you've mastered controlling the brightness of a single-colour LED by altering the duty cycle of a PWM (pulse-width modulation) signal, switching up to an RGB LED is the next logical step.

In essence, it's three LEDs in one: red, green, and blue. By combining varying brightnesses of each, you can create any shade in the RGB colour space.

You can get your RGB LED to follow a preset pattern of colours, or control it using an input device. Here, we're using the temperature reading from a basic DHT11 sensor.

First, let's wire up the RGB LED to our microcontroller – we're using Raspberry Pi Pico here, but the principle is the same. The common-cathode RGB LED has four legs: one for each colour channel and a longer leg for the ground connection. As you can see from the wiring diagram (**Figure 1**), we connect each colour channel to a GPIO pin via a 470 Ω resistor.

Once the RGB LED is wired up, it's best to check it's all working in the expected way by running a simple MicroPython program.

```
from machine import Pin, PWM
from time import sleep

# Set PWM pins to control R, G, and B LEDs
pwm11 = machine.PWM(machine.Pin(11))
pwm15 = machine.PWM(machine.Pin(15))
pwm12 = machine.PWM(machine.Pin(12))
pwm11.freq(1000)
pwm15.freq(1000)
pwm12.freq(1000)

# Loop to light R, G, B LEDs in turn
while True:
    # Red
    pwm11.duty_u16(65535)
    sleep(0.5)
    pwm11.duty_u16(0)
    sleep(0.5)
    # Green
    pwm15.duty_u16(65535)
    sleep(0.5)
    pwm15.duty_u16(0)
    sleep(0.5)
    # Blue
    pwm12.duty_u16(65535)
    sleep(0.5)
    pwm12.duty_u16(0)
    sleep(0.5)
```

Above ◈
The project wired up on a half-size breadboard. As the temperature reading from the DHT11 sensor changes, the program alters the shade of the RGB LED accordingly

Here, we start by importing the `Pin` and `PWM` methods from the `machine` library. To control timings, we import the `sleep` method from the `time` library. We then assign three GPIO pins as PWM outputs and set the frequency level (it won't work without this). Finally, we use an infinite `while True:` loop to light each colour element in turn.

Note that some RGB LEDs may have the colour channels in a different order. So, if you find yours is different, you can either change the wiring or the pin assignment in the program to swap the colour channels accordingly.

FEEL THE HEAT

Now the RGB LED is working, it's time to add the temperature sensor. We're using the basic DHT11 type here, but you could use the more accurate DHT22. The sensor has three pins: power (VCC) and ground, along with a digital output (DOUT) pin. As in **Figure 1**, we connect the former two to 3V3 and GND pins on the Raspberry Pi Pico, and the digital output to a GPIO pin. With the sensor connected, let's test it out…

```
import dht
from machine import Pin
from time import sleep

#sensor = dht.DHT22(Pin(21))
sensor = dht.DHT11(Pin(21))

while True:
  try:
    sleep(2)
    sensor.measure()
    temp = sensor.temperature()
    hum = sensor.humidity()
    print('Temperature: %3.1f C' %temp)
    print('Humidity: %3.1f %%' %hum)
  except OSError as e:
    print('Failed to read sensor.')
```

Here, we're using the **dht** library to take readings. We assign pin 21 to the DHT11 sensor (uncomment the DHT22 line instead if you want to use that). Then, in an infinite loop, we take a sensor measure and assign variables to the temperature and humidity readings, which we then print to the Shell. The temperature is in Celsius by default; to change it to Fahrenheit, you can convert it with the formula, `temp = temp * (9/5) + 32.0`. →

Right ◆

As the temperature gets higher, the RGB LED changes colour. Here it has reached red, which means it's pretty hot (for the UK) – alter the ranges in the code to suit your own location

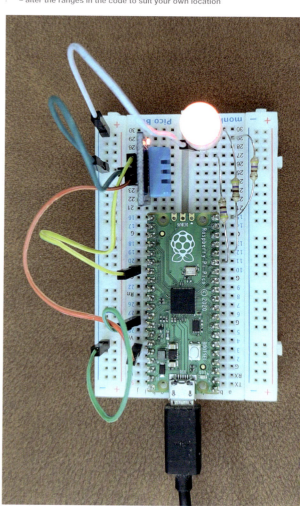

Right ◆

By varying the PWM duty cycles for the red, green, and blue channels/pins, we can alter the colour of the RGB LED to virtually any shade

TEMPERATURE RANGE

Finally, we'll use the sensor's temperature reading to set the colour of the RGB LED. The simplest way to do this would be to create a function to set a different shade for several temperature ranges. For instance:

```
def set_rgb():
    if temp < 6:
        #set LED colour to blue
        pwm11.duty_u16(0)
        pwm15.duty_u16(0)
        pwm12.duty_u16(65535)
    elif temp < 15:
        #set LED colour to yellow
        pwm11.duty_u16(40960)
        pwm15.duty_u16(24576)
        pwm12.duty_u16(0)
    elif temp < 22:
        #set LED colour to orange
        pwm11.duty_u16(57344)
        pwm15.duty_u16(8192)
        pwm12.duty_u16(0)
    else:
        #set LED colour to red
        pwm11.duty_u16(65535)
        pwm15.duty_u16(0)
        pwm12.duty_u16(0)
```

Another option is to create a colour gradient effect; with this, the LED changes shade slightly with each degree increase in temperature, going from blue, through shades of green, yellow, orange, and red. This is achieved by varying the PWM of each RGB element according to the temperature only within a certain range (or two in the case of green, so its level increases then decreases), as in the following code:

The LED changes shade slightly with each degree increase in temperature

```python
import dht
from machine import Pin, PWM
from time import sleep

#sensor = dht.DHT22(Pin(21))
sensor = dht.DHT11(Pin(21))

# Set PWM pins to control R, G, and B LEDs
pwm11 = machine.PWM(machine.Pin(11))
pwm15 = machine.PWM(machine.Pin(15))
pwm12 = machine.PWM(machine.Pin(12))
pwm11.freq(1000)
pwm15.freq(1000)
pwm12.freq(1000)

def set_rgb():
    if temp < 12:
        b = int((12-temp)*6553)
    else:
        b = 0
    if temp < 12:
        g = int(temp*5957)
    elif temp < 26:
        g = int((26-temp)*2621)
    else:
```

```python
        g = 0
    if temp > 10:
        r = int((temp)*1638)
    else:
        r = 0
    #set LED colour
    pwm11.duty_u16(r)
    pwm15.duty_u16(g)
    pwm12.duty_u16(b)

while True:
    try:
        sleep(2)
        sensor.measure()
        temp = sensor.temperature()
        print('Temperature: %3.1f C' %temp)
    except OSError as e:
        print('Failed to read sensor.')
    set_rgb()
```

We experimented and found that the conditional temperature ranges and PWM values here worked well for us, hitting a reddish shade at 26°C – considered a heatwave in the UK! You may want to alter it to better suit your locality. ◻

Tin Can Allies

Reuse before you recycle and help the environment

Above
Before 3D printers and laser cutting was so accessible, it wasn't unusual to see home electronics projects put into tin cans, biscuit tins, and other recycled containers

Dr Andrew Lewis

Dr Andrew Lewis is a specialist fabricator and maker, and is the owner of the Andrew Lewis Workshop.

The mantra of Repair, Reuse, Recycle is burned into the mind of most environmentally conscious people. We all know that the throwaway society is unsustainable, and we can do our part by keeping things out of landfill. There's a problem with this mantra, though. The first word of it pushes our thinking towards things we have that can be repaired, like clothes or high-value electronic items that we all agree deserve to be shown some care. It doesn't feel like we can move backwards and repurpose things that are supposed to be thrown away.

We take common household items for granted, and dispose of them without a second thought because they are designed to be recycled once they've been used once. Obviously, nobody is

SHARP EDGES

It's very easy to injure yourself on a piece of sheet metal, and that includes tin cans. Some can-openers leave ragged edges to the metal, and corned beef tins with a little twistkey regularly cause people to visit their local hospital. Wear leather gloves if you're working with sharp edges, and remember to double over your edges if you're making something from sheet metal.

going to try and repair a tin can and reseal the top, because that's what reusable jars are for. But, that doesn't mean we should march over to the recycle bin with our tin cans, because there is a whole world of useful projects that need the resources a discarded can provides.

Left
The indent for the motor on the side of this lathe is actually a steak-and-kidney pie tin that's been brazed into place on the case of the motor. There's no shame in reusing the tin rather than recycling it directly, and using the ready-made shape was a lot easier than making one from scratch

Below
To make a basic power pack you'll need a 4S battery controller, four 18650 batteries, a panel-mount USB socket with built-in power supply, a panel-mount cigar-lighter socket, a panel-mount voltage meter, a panel-mount fuse-holder, a power switch, and a panel-mount 2.5 mm power socket

QUICK TIP

Larger cans will fit more batteries. You could always ask a local restaurant if they have any empty catering tins you could take away. They have to pay to get their empty tins disposed of, so you'll hopefully be able to get something chunky enough to hold a lot of batteries.

POWER PACK

A portable power pack is a useful tool to have, but most of the commercial offerings are USB only, so don't give you access to the full power of the battery pack. With a few panel-mount fittings, some batteries, and a tin can, you can make a rugged power pack to suit your own specifications.

18650 batteries need a controller to make sure that they charge and discharge safely. You must use a battery welder to connect the individual cells to the controller and make a battery pack. Soldering 18650 batteries directly will damage them permanently.

Once you have a soldered battery pack, it's very easy to connect the individual panel-mount sockets and indicators to the pack via a fuse and a power switch. Be aware that 18650 batteries can dump an enormous amount of power at once, so installing a fuse is absolutely necessary and cannot be skipped. The fuse should be the first component connected to the battery pack so that every other item connects through it.

The battery controller will make sure that the battery pack charges correctly, provided that you supply sufficient voltage and current. The fully charged voltage for a 4S pack is about 16.8 V, but charging the batteries to 100% capacity will shorten their lifespan. Providing an input charge voltage of around 16.3 V should be sufficient to charge the batteries safely. You can provide this voltage from a suitably set DC buck converter. The 2.5 mm socket provides the input power point, but remember that the power socket will be wired after the power switch – so the pack won't charge unless the power switch is on.

TUTORIAL

QUICK TIP

It's best to use the
laser microscope
in a darkened room
so that the laser
projection isn't
bleached out by
natural light.

Right
The laser microscope
can be addictive,
and isn't just limited
to pond water.
Pretty much any
translucent liquid can
be examined

QUICK TIP

If you enlarge the
hole for the laser too
much, just use some
hot glue to hold the
laser in place.

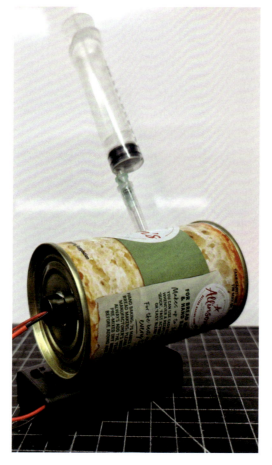

Why are tin cans so useful? For one thing, they're
a ready-made container, and they tend to come in
predictable sizes. Different cans and jars can be
connected together because they are designed to
hold a certain amount of food, and fit in a relatively
standardised industry landscape. You can glue
them, paint them, and cut, drill or punch them with
household tools.

As a source of metal, they can provide a range
of smooth or textured sheet in a variety of different
thicknesses and even different materials. For shims,
compression gaskets, and small tinsmithing projects,
they are essentially a free source of material that for
some reason we completely ignore and toss into the
recycling bin without a second thought as we wander
off towards the model shop and buy some metal
sheeting from the K&S stand. We need to accept and
normalise that just because something is designed
to be recycled, that doesn't mean it can't be reused
or repurposed.

What sort of projects really benefit from a tin can?
The obvious answers like 'paint a bean tin to make
a flower pot/desk organiser' or 'put a magnet on the
bottom of a tuna can to make a magnetic parts tray'
are functional ideas, but there's so much more you
can accomplish with a bit more thought. Essentially,
anything that goes into a project box could go into a
tin can, so let's have a look at a couple of fun ideas
that can be put together in just a few hours. If it turns
out you don't like these projects, you can still take
them apart and recycle them. ◻

QUICK TIP

Lasers are dangerous. Only use a low-powered class 1 laser for this project. Never look directly into the end of the laser emitter, and never point the laser at a surface that might reflect the laser back at you or anyone else.

Left
Some tin cans have designs printed directly onto the metal, rather than a paper label. It's surprising how resilient the printing can be. A wire wheel or some coarse wire wool and a strong solvent are probably the easiest ways to clean off the design. Alternatively, just paint directly over the top, or use a vinyl wrap

Below Right
The only things you need to make a laser microscope are a tin can with one open end, a laser emitter, a battery pack, a blunt needle in its protective sheath, a syringe, and some hot glue

> **As a source of metal, they** can provide a range of smooth or textured sheet in a variety of different thicknesses

LASER MICROSCOPE

A laser microscope is a great project for people who want to explore the tiniest parts of the world around us. It's a quick project that you can make in about half an hour. A laser microscope works by shining a laser beam into a droplet of water that's suspended from a fine needle or dropper of some sort. The droplet acts as a lens, scattering the light from the laser so that a silhouette of anything in the water droplet is projected onto the nearest wall or perpendicular surface.

Begin by drilling a hole in the bottom of your tin can, as close to dead centre as you can. The hole should be the same diameter as your laser emitter, and ideally you want it to be a push fit so that the can holds the laser in place.

Align the laser so that it is parallel with the sides of the tin can, and then make a small hole in the side of the can along the same line as the laser emitter (along the centre line of the can). This hole should be the same diameter as the protective sheath that the needle fits into. Cut 10 mm off the closed end of the sheath so that the needle pokes out of the bottom when the sheath is in place.

Push the sheath into the hole in the can, and align it so that the end of the needle is just above the path of the laser beam. You want the resting position of the needle to be set so that a droplet of water suspended at the end of the needle will be directly in the path of the laser.

Connect a battery pack with an on/off switch to your laser emitter, and put some batteries in. Glue the can on top of the battery pack, and go hunting for some dirty water with your syringe. Draw a small sample of the water into the syringe, and then insert it into the sheath that you've fitted into the can. Point the laser microscope at a plain white surface, push the syringe slightly so that a droplet forms at the end of the needle, and turn on the laser. If everything is aligned correctly, you should see an eerie phantasmagoria projected onto the wall, with tiny creatures floating in and out of focus.

TUTORIAL

More blinkies for your rucksack with NeoPXL8

Try to stay alive over winter

Ben Everard

Ben's house is slowly being taken over by 3D printers. He plans to solve this by printing an extension, once he gets enough printers.

This writer cycles to the workshop every day. At around the start of November every year it takes a dramatic turn for the dark. Gone are the glorious summer days of cycling home while the sun is still high in the sky. Gone, too, are the glorious early autumn days of the sun gently setting and lighting up the sky in hues of orange and pink. Now it's time for the dark winter days, and if there are going to be some glorious lights, we have to provide them ourselves. And provide them we will.

Here's the basic problem we're trying to solve. Bike lights are great. They let cyclists be seen in the dark. However, a single point of light isn't a great indicator of position or speed, and it requires drivers to use a degree of awareness that seems beyond some of the denizens of the road.

Bike light manufacturers have attempted to solve this problem in a few ways. Brighter lights help up to a point – beyond that just become dazzling (a particular problem on a dark bike path). Flashing lights help a little. Some lights are a little larger, and these can help significantly. However, there's

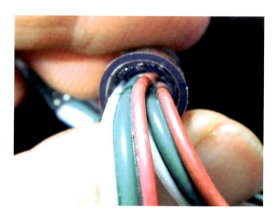

Above ◈
The electronics for each LED come encapsulated in resin. These ones have already survived being on the rucksack for a year of commuting and are still going strong

a limit to how big you can make a bike light. So, rather than attempting to enlarge a light, we've attempted to illuminate our largest item, this author's cycling rucksack.

For the last couple of years, we've had lights on the back of the rucksack. Originally, these were a simple string of battery-powered fairy lights from the sadly defunct Wilkos. Then we upgraded these with a strip of WS2811 lights (we preferred the form factor of these waterproof string lights to anything available using the smaller WS2812B LEDs). This helps us remain visible from behind, but we've never been too happy with the visibility from the front. We've tried LEDs attached to the bike, but this has always been unsatisfactory. They get knocked about more than LEDs on the rucksack, and they've never lasted long.

This year, we wanted to up our visibility by attaching LEDs to the front as well, and the only front-facing part of the rucksack – the shoulder straps.

Attaching the LEDs is only half the battle, though. We also need to control them, and fortunately, there

Right ◈
These LEDs include an external WS2811 chip and an RGB LED. WS2812B LEDs have the chip and LED in the same package. Both packages use the same communication protocol

are a couple of things that have made it easier to run multiple animations from a single microcontroller.

The first is NeoPXL8, a library that outputs data to up to eight strings of LEDs simultaneously, and LED animations, which is a module that does exactly what it says. Both of these have been developed by Adafruit and work with CircuitPython on Raspberry Pi Pico.

The first step in getting these modules to work is to download both CircuitPython for Pico and the Library Bundle from **circuitpython.org**. You can flash CircuitPython to Pico by holding down the BOOTSEL button and plugging it into your computer via USB. You can release the BOOTSEL button once it's plugged in. There should be a new USB drive that appears – drag and drop the UF2 file for CircuitPython that you've just downloaded. Once this has been copied over, the USB device should disappear and a new one called CircuitPython will appear.

Unzip the CircuitPython Library Bundle, and you need to copy across **adafruit_neopxl8** and **adafruit_led_animations** into the **lib** folder on the CircuitPython USB device. The Pico device should now be ready for our code.

There is a selection of animations available (and you can code your own). We selected Comet, which is pretty similar to a Larson scanner, and used the

GETTING STARTED
WITH CIRCUITPYTHON

If you've not used CircuitPython before, don't worry, it's all quite straightforward. We would, however, recommend that you first take a look at the getting started guide: **hsmag.cc/AdafruitCP**.

If you're an experienced programmer, the main thing you need to know is that CircuitPython is sensitive to which text editor you use. This might sound a little odd, but it has to do with the way it saves files to the device.

Mu is a great choice, and you can download it from **codewith.mu**.

Above ◈
Unfortunately, we can't capture the motion of the lights in a still image, but we're really pleased with how this looks in use

traditional red for the back and white for the front. It gives movement and light that we think is quite eye-catching, but not dazzling or distracting. We dislike lights that turn on and off sharply – we much prefer the softer introduction on this setup.

The code for all this is:

```
import board
import rainbowio
import adafruit_ticks
```

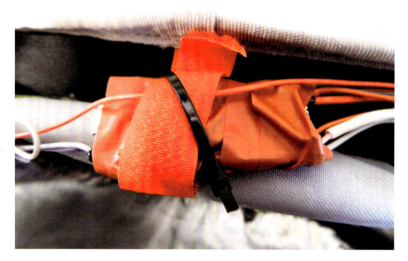

```
from adafruit_led_animation.animation.comet import
Comet
from adafruit_led_animation.color import GREEN, WHITE
from adafruit_led_animation.group import
AnimationGroup
from adafruit_led_animation.helper import PixelMap
from adafruit_neopxl8 import NeoPxl8

# Customize for your strands here
num_strands = 3
strand_length = 38
front_len = 6
first_led_pin = board.GP2

num_pixels = num_strands * strand_length

# Make the object to control the pixels
pixels = NeoPxl8(
    first_led_pin,
    num_pixels,
    num_strands=num_strands,
    auto_write=False,
    brightness=0.50,
)
```

OTHER HARDWARE CHOICES

Raspberry Pi Pico is a great bit of hardware for this, but it's not the only option. Any board with an RP2040 processor and CircuitPython support should work. Adafruit has a board that's purpose-built for controlling lots of LEDs: the Scorpio. This simplifies the wiring a bit and includes a LiPo battery charger if you don't want to use a USB battery pack.

```
def strand(n):
    return PixelMap(
        pixels,
        range(n * strand_length, (n + 1) * strand_
length),
        individual_pixels=True,
    )

strands = [PixelMap(
        pixels,
        range(0,37),
        individual_pixels=True),
    PixelMap(
        pixels,
        range(38,44),
        individual_pixels=True),
    PixelMap(
        pixels,
        range(76,83),
        individual_pixels=True)
    ]

# For each strand, create a comet animation of a
different color
animations = [
    Comet(strands[0], speed=0.02, color=RED, tail_
length=10, bounce=True),
    Comet(strands[1], speed=0.05, color=WHITE, tail_
length=3, bounce=True),
    Comet(strands[2], speed=0.05, color=WHITE, tail_
length=3, bounce=True),
]

# Advance the animations by varying amounts so that
they become staggered
for i, animation in enumerate(animations):
    animation._tail_start = 30 * 5 * i // 8  #
pylint: disable=protected-access

# Group them so we can run them all at once
animations = AnimationGroup(*animations)

# Run the animations and report on the speed in frame
per second
t0 = adafruit_ticks.ticks_ms()
frame_count = 0
while True:
    animations.animate()
    frame_count += 1
    t1 = adafruit_ticks.ticks_ms()
    dt = adafruit_ticks.ticks_diff(t1, t0)
    if dt > 1000:
        print(f"{frame_count * 1000/dt:.1f}fps")
```

```
    t0 = t1
    frame_count = 0
```

This is almost the example code for NeoPXL8 (Thanks, Jeff Epler), but with a few changes to make it work how we want.

The `pixels` object is very similar to one created with the usual NeoPixel library. It's a one-dimensional list with the first strip taking up places 0–37 (in our case), then the second 38–75, etc. The slight confusion here is that our strips aren't all the same length. We have a single strip of 38 pixels, then two strips of six pixels. The NeoPXL8 library can't handle this, so it sets every strip up as 38 pixels and we have 32 'phantom' pixels in the second and third strips. This doesn't cause us any problems – we just have to ignore them.

We don't want to address all our pixels as one list, and we want to interact with the three strips individually. Fortunately, there's the `PixelMap` object for this. We can use this to divide up our `pixels` object into the three strips (and we can also discard the phantom pixels). This is what the strands list is – a list of three objects, each of which is an LED strip.

That's the setup done; let's now start with the animations. The LED Animations module contains a series of different animations (as well as a framework you can use for building your own animations). In our case, we've used three Comet animations.

The three animations can be bundled together into an animations group. This makes it a bit easier to update and run them. Finally, we just need to repeatedly call the `animate` method of the group repeatedly, and this will run our animations.

LET'S BUILD

That's the code. Now let's take a look at the physical setup.

Like many rucksacks aimed at people who do outdoor activities, this one comes with a wide range of superfluous bits sewn into it. We're not sure exactly how the good people of Berghaus intend people to use the ice axe loops that are too small for any ice axe we've ever seen, but they are handy for attaching LEDs to. We cable-tied the LED chain to every extraneous bit we could see. It's in a bit of a jumble – this is intended as it gives more of a 'cluster of light' effect than a more mathematical layout might.

As mentioned, there are three separate strips, each of which is wired to a different pin. We have to solder them all up. You can use whatever GPIO pins you like, but they must be consecutive (we used 2, 3, and 4). Pico has enough ground pins, so we don't have to do any creative wiring there, but we do have to connect

the VIN pins of each strip together and then to the single VBUS pin. We joined all three wires together with a big drop of solder and a lot of electrical tape (we don't have heat-shrink the right size for this).

The main thing you need to think about with construction is strain relief and minimising the likelihood of tangles. We want this to last a long time, and it's going to get knocked about as bags tend to do. The main thing we've used is cable ties. A small cable tie through the mounting hole on Pico can hold wires in place and stop them from flexing at the soldered joint (which is the most likely place to fail). Cable-tying wires together and in place as much as possible will help keep things tidy (and therefore safe and working).

It remains to be seen if your setup will prove robust enough. There's definitely more bending on the new front wires than there has been on the back. That's just an unavoidable part of where they're placed. If it breaks, we'll fix it and introduce a bit more robustness then.

Finally, we need a power supply, and for this, we just use a USB power bank in the rucksack.

We're happy with the new setup, and it seems the good people of the West of England are, too. We get regular comments from fellow cyclists on our look. □

Left ◆
A cable tie through a mounting hole isn't the best form of strain relief, but it's much better than no strain relief

HackSpace
TECHNOLOGY IN YOUR HANDS

WEEKEND PROJECTS

HACK | MAKE | BUILD | CREATE

When you've got a bit more time to yourself and can spend longer in the shed/garage/workshop, these brilliant builds will keep you occupied and expand your making skills

PG 30

3D PRINTING

Print in Polyvinyl Butyral (PVB) and dissolve layer lines for a smooth finish

30

PG 36

3D PRINTED LITHOGRAPHS

Paint a picture with layers of 3D printed colours

PG 42

GIGABUTTON

Combine 3D printing, coding and electronics to build a great big button

42

GO TO PRESS

TUTORIAL

Polyvinyl butyral (PVB)

Get glassy-smooth prints with PVB and alcohol

Ben Everard

Ben's house is slowly being taken over by 3D printers. He plans to solve this by printing an extension, once he gets enough printers.

Solvent smoothing is one of those things that seems great in theory, but can be a pain in reality. The idea is simple. All 3D prints have layer lines, but you can remove them by dissolving the outer layer of plastic in a small amount of solvent. The now-liquid plastic will flow into the crevices between the layers, and the solvent will evaporate, leaving the plastic that was dissolved in it to fill any gaps and produce a shiny surface.

The main problem with it is that it usually involves harsh chemicals. PLA, in particular, needs substances like chloroform, which requires equipment beyond our workshop setup to do it safely, so we've been unable to test it out. ABS is a little more straightforward as it can be smoothed with acetone. However, even this is still unpleasant to work with.

PVB is soluble in isopropyl alcohol (IPA). While there are still issues to be considered when working with IPA, it's pretty safe to assume that most people who 3D print are comfortable using this particular chemical. However, don't assume that you can easily do this safely because you're used to using IPA for other things. Take a look at the Safety box (opposite), and make sure that you're properly considering your own experience and equipment.

Solvent smoothing aside, the only other reason we can think of to use PVB is because it's transparent. It's a little clearer than other transparent filaments we've seen, but only slightly. Unless you need absolute maximum transparency, you will probably find it easier to use transparent PLA.

ABSOLUTE MELT

That's the theory; let's see how it works in practice. PVB is reasonably easy to print. It requires a slightly hotter bed than PLA, but it's nothing that will cause problems for most modern printers.

The biggest problem with printing PVB is that it is very hygroscopic, meaning that it absorbs moisture from the atmosphere. If you plan on printing it quickly, you might be able to get away with keeping the filament in a sealed bag containing a desiccant sachet. However, ideally, you should have a filament dryer.

When printing something that's going to be transparent, as most PVB prints will be, you need to consider the inside of the print – in other words,

Right ◆
A Benchy before smoothing. The transparent filament does still look good, but has layer lines as usual

POLYSHER

Polymaker makes a device specifically for smoothing PVB objects with IPA called the Polysher. We've not tested this out, so we can't comment on how well it works, but it should make the process a little more straightforward and less messy.

Left ◆
After smoothing, the
layers are still visible
to the eye, but the
surface is smooth
to the touch

 When printing something
that's going to be transparent,
**you need to consider the
inside of the print**

the infill. This pattern can ruin the look of your
transparent object. Some prints you can make
completely hollow (such as with Vase mode), but
if this isn't possible, you can completely fill it. You
can do this by adding 100% infill or by upping the
number of perimeters to something far bigger than
would actually fit (such as 999). We'd recommend
the latter because it will lay down the filament
concentric with the outside perimeter, which is less
likely to produce visible issues. Obviously, filling up
an object with plastic is going to have a pretty big
negative impact on both the amount of filament that
it uses and the amount of time it takes to print. →

SAFETY

It's easy to get complacent about a chemical you use
regularly, but it's important to remember the dangers
of IPA because when applying it with a brush, you're
more likely to have a problem. We recommend reading
the material safety data sheet (this should be available
from the supplier). Particular risks are:

- Inhalation
- Eye contact
- Fire

All of these are significantly more likely when
brushing it on a model than when wiping it on a print
bed. Use proper eye protection. It's recommended that
you use nitrile gloves when handling IPA.

Remember that safety is your responsibility, and you
must fully understand the risks associated with a
process before starting it. Proper PPE is a requirement
for this process, not an optional add-on.

Above ↗
In Vase mode, there's
no infill to see, so the
transparency isn't
disturbed. This photo
is before smoothing

If you have a sufficient flair for 3D design, you could probably make something interesting using the voids inside the print.

There are two main brands of PVB printer filament: Prusament PVB and Polymaker PolySmooth. We tested out Prusament, but the same process should work with PolySmooth.

Print out your part as you would normally. Since we're going to be smoothing this, there's no need to worry too much about layer lines, so go big for a fast print. This might cause problems with small details,

but smoothing can cause problems with small details anyway, so this process is best suited to objects that can be printed with a large layer height. If you've got a big nozzle, this can be a good time to use it.

Once you've got your model, you need to apply the IPA to the surface. It should be at least 70% IPA. You can do it in a few ways:

- immerse it
- squirt it
- brush it
- vaporise it by applying it to a cloth; putting both cloth and 3D print in an enclosed space, such as Tupperware

Whichever way you do it, the key is to get a uniform coating, and leave it to dry. When drying, you want to place it on something that minimises the number of contact points, as these will leave blemishes on the surface. In most cases, you'll need to repeat the process several times in order to get the desired effect.

We tested this out by brushing on IPA – this is probably the most approachable option for people trying this out. The method is really simple. Take

VASE MODE

You can get great results with PVB in Vase mode. The resulting prints are almost see-through and have a glass-like shine. However, you do need to be careful because the walls are very thin. The vases can become very weak. We accidentally dissolved a hole in one of the test pieces.

While with Vase mode, you can only have one perimeter, this perimeter can be very thick. By default, PrusaSlicer sets the external perimeter wall to 0.45 mm on a standard 0.40 mm nozzle, but you can adjust this. We upped it to 1.00 mm and had good results.

Left ◈
We heavily smoothed
this one and the
layers are only just
visible. However, it
has distorted the vase
slightly. Corners are
more rounded and
finer detail is lost

 There's something very
appealing about the
**transparent, shiny prints
you can get with PVB**

your print, a paintbrush, and a small amount of IPA.
Paint the IPA onto the print and try to get as smooth
a coat as possible. You need to be quite liberal with
the IPA to get a good effect. You'll find that the IPA
will spread itself out along the layer lines. We'd
recommend using nitrile gloves when doing this.

Once you've got a good coat, leave the object to
dry on a suitable surface – this will take about half
an hour. If the object you're making is hollow – like
a vase – you can swish some IP around inside to
get an even coat. Pour out any excess and leave
to dry upside down. It can take about five to ten
applications of IPA to smooth away the layer lines.

RESULTS

We are quite partial to the look of layer lines, and
they're not something we generally feel the need
to sand or post-process away. That said, there's
something very appealing about the transparent,
shiny prints you can get with PVB. When done well,
it can be almost glass-like.

There are, however, downsides. The process
that removes layer lines will also remove other fine
details (just as sanding, painting, and other post-
processing would). It's a bit tricky to apply the IPA
smoothly and evenly.

While Vase mode does work well, it does produce
parts with thin walls, and the IPA will remove
material from some parts – this can leave your print
weak in places.

There are certainly prints that can look excellent
when printed in PVB and smoothed. Things with
gentle curves that are prone to false contours can
smooth particularly well. ◻

SUBSCRIBE TODAY
FOR JUST £10

Get **3 issues** + **FREE** Pico 2 W

SUBSCRIBER BENEFITS

Free Delivery
Get it fast and for free

Exclusive offers
Great gifts, offers, and discounts

Great savings
Save up to 37% compared to stores

SUBSCRIBE FOR £10

Free Pico 2 / Pico 2 W

3 issues of Raspberry Pi Official Magazine

£10 (UK only)

SUBSCRIBE FOR 6 MONTHS

Free Pico 2 / Pico 2 W

6 issues of Raspberry Pi Official Magazine

£30 (UK)	$43 (USA)
€43 (EU)	£45 (Rest of World)

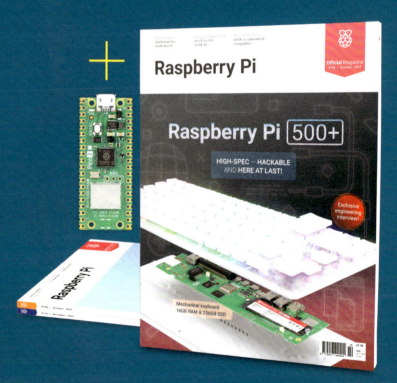

📞 Subscribe by phone: 01293 312193

📨 Subscribe online: rpimag.co/subscribe

✉ Email: **raspberrypi@subscriptionhelpline.co.uk**

SUBSCRIBE TODAY AND GET A

FREE Raspberry Pi Pico 2 W (or Pico 2)

Subscribe in print today and get a FREE development board

- A brand new RP2040-based Raspberry Pi Pico 2 W / Pico 2 development board
- Learn to code with electronics and build your own projects
- Make your own home automation projects, handheld consoles, tiny robots, and much, much more

WORTH $7

Free Pico 2 or Pico 2 W. Accessories not included. This is a limited offer. Not included with renewals. Offer subject to change or withdrawal at any time.

➤ Buy now: **rpimag.co/subscribe**

SUBSCRIBE on app stores

From **£2.29**

Available on the App Store

GET IT ON Google Play

TUTORIAL

Adventures in multicoloured printing

Use your plastic squirter to make impressive pictures

Rob Miles

Rob has been playing
with hardware and
software since almost
before there was
hardware and software.
You can find out more
about his so-called life at
robmiles.com.

In this article, we are going to look at how
you can add colour to your 3D prints.
We'll start with a look at lithophanes and
them move onto tools you can use to create
impressive 3D artwork. You can find the
source images and the 3D printable files for
this article here: **hsmag.cc/Colour3D**.

LITHOPHANES

Lithophanes are a great answer to the frequently
asked question: 'Why did you get a 3D printer?' You
can reply: 'To make these' and hold a thin piece of
printed plastic up to the light to reveal a neat image.
The next question that is asked is usually: 'Can you
make me one of those?'

Figure 1 below shows a lithophane panel made
by the author a few years ago of his then-new
car. On the left, you can see the surface of the
3D-printed panel and, on the right, you can see how
the lithophane looks when light is shone through
it. If you look at the panel on the left, you can see

that the dark areas of the image are thicker than the
lighter ones. It's funny how such a low-quality print
can look so good as an image.

MAKING YOUR OWN LITHOPHANES

If you have a 3D printer, you should have a go at
printing at least one lithophane. They make nice
personalised gifts. Lithophane panels are created by
software that converts light levels in an image into
thickness values in a 3D-printed panel. The darker
the image, the thicker the panel, so that less light
gets through. You can create curved or spherical
lithophanes which you can make into great lights.
The Lithophane Maker at **lithophanemaker.com**
is a good place to start. You load an image into the
web page and the software generates an STL (an
abbreviation of STereoLithography) file which you
then slice and print in white PLA. There are also
designs available for lights into which the lithophane
can be mounted. We'll now consider a few things
that you need to remember.

Figure 1 ⊠
This lithophane was
printed on standard
white PLA filament
using a Creality Ender
3. It took a few hours
to complete

YOU'LL NEED

- **A 3D printer**
(preferably one
that can switch
between different
filaments)

- **For coloured
lithophanes**
you will need some
cyan, magenta,
yellow, and
white filament

- **For drawing
plaques**
you can use
any filaments
you happen to
have around

Figure 2
The selector has been loaded with cyan, magenta, yellow, and white filaments to make a colour lithophane

PRINTER **POOP**

Each time the printer switches to a different filament colour, it purges the previous colour out of the print head and replaces it with the new one. The remnants of each colour change are ejected from the back of the printer so, after printing for a while, you get a little pile of 'printer poop'. These plastic fragments give you an interesting record of your printing history and are not without artistic merit, as the picture above shows.

LAYERS AND NOZZLES

A 3D print is created as a series of layers stacked on top of each other. The printer is sent a file telling it where to put the filament that makes up each layer. The printer file is produced by a program called a 'slicer' which takes an object design and converts it into a series of horizontal slices (or layers) to be printed. When a design is 'sliced', you need to tell the printer the 'layer height' which gives the height of each slice to be printed. When slicing a lithophane,

> **The darker the image,
> the thicker the panel**
> so that less light
> gets through

the smaller the layer height, the more levels of 'grey' you will get in the image. In **Figure 1**, you can see noticeable banding in the sky where the brightness changes from light to dark. If I had used a lower layer height for the print, this banding would have been less pronounced. Most printers regard 'normal quality' as 0.2 mm layer height (five layers per mm), but for a better-looking lithophane, you can reduce the height to 0.1 mm – although this will double the time the lithophane takes to print as the printer must now produce twice as many layers.
You can also improve the resolution of the image by reducing the size of the printer nozzle. The nozzle

is the hole where the extruded plastic comes out of the printer to be added to the printed object. Most consumer 3D printers have a nozzle which is 0.4 mm in diameter, but, with many models, you can switch the nozzle for one which is smaller – usually 0.2 mm. This lets the printer produce more detailed objects (useful for things like gears and the like) at the expense of a huge increase in printing time. The author has found that he is quite happy with the images produced by the 0.4 mm nozzle. There are some samples of each size later in this article.

DIGGING INFILL

A slicer has an 'infill' setting that determines the amount of filament extruded inside the object being printed. It is expressed as a percentage. An object printed with an infill of 0% would be completely hollow (and not very strong). An object printed with an infill of 100% would be completely solid. The normal setting for infill is 20%. This is fine for most objects unless we want them to be particularly robust. For a lithophane, we need the interior areas to be solid to stop all the light, so the infill for the slicer should be set to 100%. This also has the effect of increasing the printing time.

COLOUR LITHOPHANES

The first 3D printers printed one colour. You could only print different colours by swapping the filament during the print. However, today, you can get printers which can automatically switch between filaments during a print job. **Figure 2** above shows the AMS (automatic media selector) used with the author's Bambu Lab P1S printer (**bambulab.com**).

QUICK TIP

The first lithophanes were produced in the 18th century. They were made from porcelain. An artist would carve a 'master' image in wax that would be used to make moulds into which liquid porcelain (a type of clay) would be poured. When the porcelain hardens and is fired in a kiln, it becomes translucent, with the thicker areas stopping more light, just like 3D-printed lithophanes.

Figure 3 ◈
You can preview the expected image on the web page, but this view is not as impressive as the final printed one will be

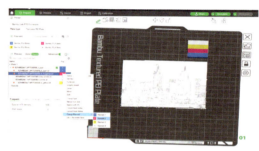

Figure 5 ◈
The magenta filament is being selected for the magenta layer of the print

The AMS is connected to the printer which tells the AMS which filament is required during the printing process. When the printer needs to change filaments, the current filament is wound back onto the roll and the new one is loaded into the print head. The printer then pushes the new filament through the head to purge out any remaining pieces of the previous filament before resuming printing with the new colour.

If your printer can print multiple colours (or you are prepared to manually switch filaments during the print job) you can make coloured lithophanes. These work in the same way as black and white ones, but instead of a single layer representing the light and dark areas of the image, you now use four colours: cyan, magenta, yellow, and white.

Producing the lithophane files is the same as for monochrome, although the web address for the generator site is very slightly different: **hsmag.cc/ColourLithophane**.

Figure 3 shows an image being prepared for conversion into a lithophane. The image being used to create the lithophane is a slightly processed photograph of the Melbourne skyline. The output from the generator is a zip archive which contains six STL files: two for white (the bottom and the top of the lithophane) and one each for the cyan, magenta, and yellow layers.

Figure 4 shows how the layers work together to create a coloured picture. The colour layers are sandwiched between the two white layers.

Figure 4 ◰
In the actual lithophane object, these images are all directly on top of each other. In the picture, they've been separated so that we can see how they work

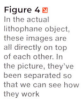

QUICK TIP

Using a media selector makes it easier to print right up to the end of a roll of filament. You can load a full roll into the selector and tell the printer to switch to that when the current roll runs out.

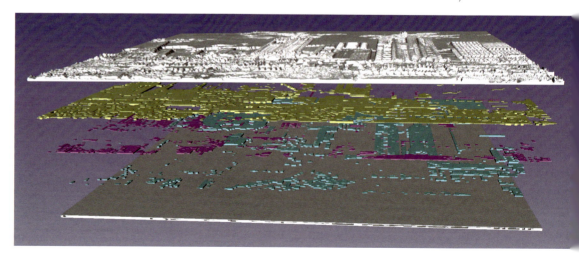

Figure 6 ◈
This print took less than two hours to complete

Figure 7 ◈
This print took over eight hours to complete

Figure 8 ◈
There are also tools in Bambu Studio that can embed text into objects and make coloured signs

Light travelling through the layers is filtered so that only the required colours are visible. The STL files are loaded into the slicer program which will produce the control file for the printer. The author uses the Bambu Studio slicer program. This can produce print files for a wide range of different 3D printers, not just ones made by Bambu Lab. You can download it from here: **hsmag.cc/BambuLab**.

You can improve the resolution of the image
by reducing the size of the printer nozzle

Figure 5 shows the Melbourne lithophane being prepared for slicing. Each colour layer STL has been imported as one element of a composite object. This means that the layer objects will be printed on top of each other, rather than being distributed around the print bed as different objects. The printer control file that is produced will contain

commands to the printer to switch filaments at the appropriate times.

LITHOPHANE QUALITY
Figure 6 shows the kind of results you can expect if you use standard printer settings. There is a reasonable amount of detail, but the banding on the sky is quite pronounced.

Figure 7 shows the results achieved by reducing the layer height and using a 0.2 mm nozzle. The sky looks a lot better and there is much more detail, particularly in the buildings on the right of the picture. Whether the improvement in detail is worth an extra six hours of print time is something worth thinking about though.

COLOURED PRINTS
Lithophanes are great fun and produce very impressive results, but they are not the only way of making coloured prints. Lithophanes also need a light behind them, which restricts where you can use them. However, we can also use a 3D printer to produce coloured prints simply by taking an existing print and 'colouring it in'.

Figure 8 shows how the Bambu Studio paint tool lets you use the mouse to paint primitive messages into a 3D model. You can also use the tool to select parts of an object and colourise them – for example, →

QUICK TIP

When adding colour to a print, remember that filaments may have to be changed in many layers. The slicer will usually tell you how many filament changes are required for a given print.
You can greatly reduce print time by adjusting the colours in your design to minimise the number of colour changes.

Adventures in Multi-Coloured Printing

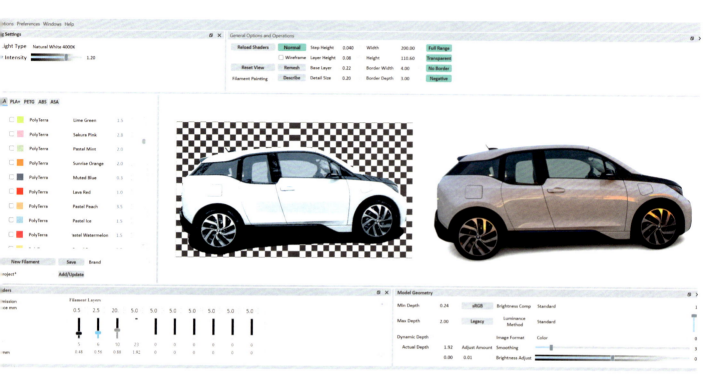

you could colour in features of an animal model. When the object is saved from the slicer, the colour changes are saved as well, and when the object is sliced, it will contain colour change commands to be used during printing.

Object painting is useful for adding colour to existing models, but it would also be nice to be able to take an image and prepare it to be 'rendered' by a 3D printer as an object. It turns out that there is a tool available to do just that, and it is called HueForge.

HUEFORGE

HueForge is a tool that takes 2D images and makes them into 3D-printable plaques. You can find it here: **shop.thehueforge.com**. **Figure 9** shows it being used to create a printable car picture. The program imports a colour image and maps colour or intensity values onto filaments that will be used to create a 3D-printable picture. You select the filament colours you want to use and drag them onto the sliders at the bottom left of the program. In **Figure 9**, the filament colours have been set to black, blue, grey, and white, which were loaded into the printer before the print. You can see these sliders at the bottom left of **Figure 9**.

You adjust the sliders for each filament to select the light intensity level in the source image that you want to map that filament colour onto.

As you can see above, the black filament has been mapped onto the darker colour, the blue onto a range of colours that match the blue parts of the image, and the grey and white filaments onto brighter parts. HueForge produces an STL file which describes the object to be printed. Before you slice the object, you need to configure the slicer to change the filament at particular layers. HueForge creates a text file which tells you what to do:

```
Swap Instructions:
        Start with Black
        At layer #5 (0.48mm) swap to Aqua Blue
        At layer #6 (0.56mm) swap to Silver Smoke
Grey
        At layer #10 (0.88mm) swap to Jade White
for the rest.
```

Above, you can see the layer colour instructions for the car picture. The first four layers are black, followed by blue, grey, and white. Bambu Studio allows you to select a layer and specify the filament colour for that layer.

You can have a lot of fun experimenting with different filament colours and adjusting the sliders to generate different 'looks' for your finished image. HueForge works even better if you spend some time preparing an image to be processed.

Figure 9 ◈
The background removal tool at 'remove.bg' was used to remove the car from the source picture

QUICK TIP

You might think that you would use red, green, and blue filaments for colour lithophanes, but, like printing on paper, the print will work by filtering out the colours we don't want rather than lighting up the colours we do.

The author of HueForge has made some useful videos exploring how to do this. You can find them here: **hsmag.cc/HueForgeFAQ**.

When adding colour to a print, remember that filaments may have to be changed in many layers

Figure 10 shows the first HueForge print made by the author. It turned out very well and shows the level of detail that can be achieved. There is a very impressive 3D effect on the door handles and the panel lines on the car. With a bit more work, it is hoped that the blue elements could be made to stand out more.

It turns out that a 3D printer is not just for boxes and mechanisms but also has artistic merit too. **Figure 11** shows the kind of fun you can have. Modern printers can print large, flat objects much more quickly than previous ones, so you can turn out artwork quickly. The author hopes you have as much fun doing this as he has. □

Figure 11
This print uses just three colours: black, white, and grey

Figure 10 ◈
The panel at the top right of the print is the 'purge tower'. This is used by the printer to make sure that all of a previous colour is removed from the print nozzle before it is used on the print.

QUICK TIP

By picking a source image with little detail and large areas of solid colour, you will find that you can produce good-looking colour lithophanes even with low-resolution printer settings.

TUTORIAL

Big internet-connected button

One whack to send the magazine to press

Ben Everard

Ben likes buttons of all sorts, but his favourite are ones big enough to hit with his whole hand. He's very much looking forward to finishing this issue so he can whack this one.

Each month at Raspberry Pi, we create a magazine. We source content, edit it, massage it into the right format, add images, and check through everything with a fine-toothed comb. Once this is all done, we send it to the printers to translate our digital files onto paper.

It's this final step that feels a little, well, underwhelming. You might imagine that sending a magazine to press involves hitting a big important-looking button. At least, that's what this writer thought when he first joined the world of publishing many years ago. However, reality is less impressive, and we actually just send a message to someone to say that we're happy with it.

When reality is underwhelming, we can choose to accept that, or we can choose to enliven it. This issue, we choose the latter and we've created our own big, important-looking button. OK, technically we haven't created our own button, we bought it (the aptly named Massive Arcade Button 100 mm from Adafruit). We did, however, make a mount for it, attach it to a Raspberry Pi Pico W, and program it to send the appropriate message when we hit it.

Now, rather than type out our approval message, we just hammer the big red button once the issue's ready to send.

Obviously, it's unlikely that you happen to need a button to send a magazine, but we'll look at how to make a big red button that does – well, anything you like really.

First, you'll need the 3D-printable file for the box. You can grab this from **hsmag.cc/button_box**. It's designed to work with a multi-colour printer, but you can print it in a single colour. It's a pretty simple design – just a cube with the unnecessary bits removed, leaving a hole in the top for a button, and a hole in the back for power. There's no mounting for the Raspberry Pi Pico W inside – we prefer to use a bit of hot glue but, if you're averse to that sort of thing, you could add a few holes.

The text was added in PrusaSlicer, and can be edited in that software as well.

The electronics for the switch are in a removable assembly in the bottom of the button. Twist it a few degrees and it should pop off. There's an LED in the button, so we wire that up to a GPIO pin (at present, we just leave it on, so we could wire it up to 3.3 V

ALTERNATIVE BUTTONS

We used a massive arcade button for this, but there are plenty of other options. At its heart, the button is just a microswitch, and you could set many different things to also trigger the microswitch. Alternatively, in CircuitPython, you can create touch sensors on Pico. Find details at: **hsmag.cc/ rp2040_cap_touch**. Using this, you could create a pad out of something conductive that sends a message when you touch it. Perhaps a tinfoil hand that triggers the message when you high-five it?

When reality is underwhelming, we can choose to accept that, or we can choose to enliven it

or VBUS, but we thought we'd keep the option for adding light effects in the future). The LED has a built-in resistor, so we can just connect it up directly to pin 0 and ground. The LED pins are on either side of the switch (see **Figure 1**). You can use either one as ground and power, and then just turn the LED around to match.

There are three pins for the switch: Common, Normally Open, and Normally Closed. We wired up Normally Open and Common directly between GPIO 1 and ground. The Common pin is on the bottom of the unit, while the Normally Open is the bottom of the two pins on the end. We can then add a pull-up resistor in software, and the button will read 1 when it's not pressed and 0 when it is.

That's the extent of the hardware for this build, let's now take a look at the software.

We've used MicroPython, so you'll need to install that to the Pico W.

We need to send our message via Slack, and this has an application programming interface (API) that lets us interact with it from our script. If you want to work with a different service, you need to find out the suitable details of its API.

DIALLING IN

Interacting with the Slack API is a little fiddly because you have to get the security details right. Fortunately, someone far smarter than us has already done the hard work and written it up on the Raspberry Pi blog. There are full details at **hsmag.cc/slackbot**. We won't rehash all of this here. The most important part, for our purposes, is setting up the security details on the Slack website (depending on your configuration, you may need to get your workspace admin to approve the bot).

Once you've set all this up, and noted down the security details, you need to get the code from: **hsmag.cc/git_slack**.

The key files we need are in the MicroPython directory. We need all of these except **main.py**, for which we'll use our own code (below). These need to be copied onto the Pico W, and you can do this using the Thonny MicroPython editor. Go to View > Files, then you can navigate to the place you downloaded them. Select them, and upload them to the **/** directory on Pico. →

Left ◆
The combination of different coloured letters, plus a slight embossing, really helps the letters stand out

Below ◆
There are few things quite as exciting as a massive red button that's just begging to be pressed

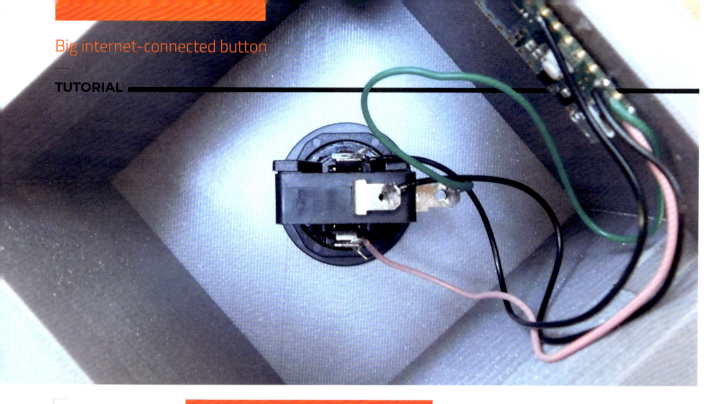

Above ◈
Figure 1. The button
has five connections
– we use four of them

TYPE IT OUT

If you want to interact with a service that doesn't
have an API you can use, you can get Pico to connect
to your computer as a keyboard and type out a
message when it's pressed.

At the time of writing, support for this in
MicroPython is incomplete, but you can get the code
from here: **hsmag.cc/micropyton_hid**.

Alternatively, you can use CircuitPython, which
supports Pico and has full support for emulating a
keyboard. You can find out more at this link:
hsmag.cc/circuitpython_hid.

SIMPLE SCRIPT

Let's now look at our own code (below).

The one bit that you have to change here is the
`channel_id`. In Slack, you can send a message to
a person, a group of people, or a channel, but all
three options use a channel ID. To find out the
channel ID of a channel, click on the channel, then
the little drop-down arrow next to the channel
name, and scroll to the bottom of the box that pops
up, and you should see the ID.

For a message to one person, go to that person's
profile, click on the three dots, and you should get
the option to copy the member ID. This can be used
in place of a channel ID.

For a group of people, you have to already have a
chat with the group going. Click on the conversation,
and then there's a drop-down arrow next to the title
(which is the names of the people in the chat). Click
on this, then go to the About tab and you should see
the channel ID at the bottom.

The code is pretty basic. First, it turns on the LED,
and connects to the network. Once it's connected, it
starts to loop, waiting for the button to be pressed.
When it is, the code sends the Slack message.

We need to be a little careful that, with each
press of the button, we only send the message
once. When a button is pressed, two contacts come
together and, as they do, they can connect and
disconnect multiple times before settling together.
This is known as bouncing.

We have a little loop that runs to make sure that
the button has been fully unpressed before it can
fire again.

The full code for this is:

```
import network
import time
import config
from slack_bot import SlackBot

channel_id = "put your channel here"
message = "Let's send it"

led = machine.Pin(0, machine.Pin.OUT)
led.on() # let's just have it light up
button = machine.Pin(1, machine.Pin.IN, machine.
Pin.PULL_UP)

# initialize the Wi-Fi interface
wlan = network.WLAN(network.STA_IF)

# activate and connect to the Wi-Fi network:
wlan.active(True)
wlan.connect(config.WIFI_SSID, config.WIFI_PASSWORD)

while not wlan.isconnected():
    time.sleep(0.5)

print(f"Connected to Wi-Fi SSID: {config.WIFI_
SSID}")
print("Now waiting for the button")
while True:
    if not button.value():
```

```
        #The button has been pressed
        slack_bot = SlackBot(config.SLACK_APP_
TOKEN, config.SLACK_BOT_TOKEN)
        print("posting message")
        slack_bot.post_message(message, channel_
id)
        print("posted message")

        #wait for button to be unpressed
        counter = 0
        while True:
            if button.value():
                counter += 1
            if not button.value():
                counter -= 1
            if counter > 10:
                break
            time.sleep(0.01)
```

Load the code onto Pico W (along with all the libraries from the original project).

Make sure that you have the configuration file set up, and you should be able to send messages with the press of a button. Whether it is to send something to manufacture, share a joke, or anything else you might need, whack the button to send the message. ◻

MESSAGE **BACK**

Our button simply sends messages to Slack, but the API allows messages to be sent the other way as well. Take a look at the original link for more details.

Depending on what you want to do, you could send a message to 'arm' your button before it's used, or light it up. You could add servos or colourful LEDs to display some information when it's ready to go.

Above ◪
A spot of glue mounts Pico W

Below ◪
Everything's good – let's go to press!

Don't forget the batteries!

How to power a single-board computer or microcontroller with a battery pack

Phil King

Phil King is a freelance writer and sub-editor specialising in technology. The former books editor for Raspberry Pi, he is also the author of *Chromebooks In Easy Steps.*

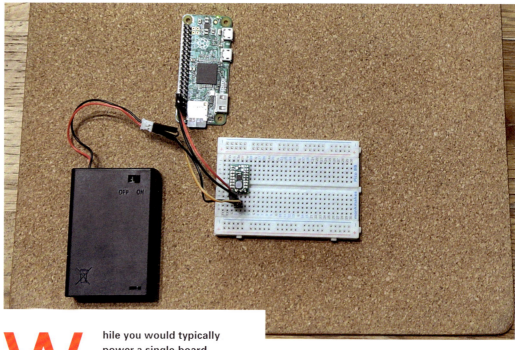

While you would typically power a single-board computer or microcontroller through its USB socket (whatever the type), it is also possible to power most boards from a source wired directly to their 5 V power and GND (ground) pins. This includes all the Raspberry Pi SBC models, along with microcontrollers such as Arduino, Pico, and ESP32/ESP8266 boards. That's what we are going to explore in this tutorial, looking at the pros and cons, including the risks – you don't want to see that magic smoke!

WHY DO IT?
Depending on your project, there may be various reasons why you would want to power the main controller board directly via the pins instead of the main USB power port.

A common use case is when you want to supply power from some sort of battery source, which could be standard alkaline or rechargeable NiMH batteries (typically AA size), Li-ion cells, or a LiPo battery (see 'Battery types' box, overleaf).

For instance, you can power a Raspberry Pi SBC using a power HAT or SHIM to connect a LiPo battery. You could even use a coin cell if space is really tight.

The direct-to-pins route may also be appropriate if you want to power multiple parts of a project, such as motors and servos in a robot, from a single source – which could be higher voltage than you need for the board, in which case you'll need a buck (or step-down) converter to regulate the voltage. We'll take a look at some options for that later, but first we're going to do the 'wrong thing'!

Above Right
It's advisable to use a voltage regulator between the power source and board. Here, we're using a Pololu S7V8F5 to maintain a steady output voltage of 5 V to a Raspberry Pi Zero

DON'T TRY THIS AT HOME

Yes, in most cases, you really can just wire up a suitably rated battery source to your board's 5 V and GND pins to power it. This basic method is not without its risks, however.

For one, unlike when using a USB power port, there is typically no power regulation or fuse protection on the GPIO pins of a board such as a Raspberry Pi to protect it from overvoltage or current spikes which may damage or even destroy it. You have been warned! You also need to ensure you connect the pins the correct way around – reverse polarity is not good.

Still, for starters, we're going to try powering a Raspberry Pi Pico microcontroller and then a Raspberry Pi Zero SBC using this method, just to check that it works.

The lower power requirements of these two boards do make it easier to supply them from a battery source, compared to larger, more power-hungry boards like the Raspberry Pi 4 and 5.

Using a standard 3 × AA battery holder (available from most electronics suppliers) with three 1.5 V alkaline batteries, we used a digital multimeter to measure the output voltage at a shade under 4.5 V, but this may vary according to the newness and type of batteries. If you use rechargeable NiMH batteries, you may well find the voltage is lower, around 3.6 V – so if you really want 5 V, you'll need to use a 4 × AA battery pack instead (don't do that with alkaline ones, unless you're using a buck converter to reduce the voltage).

We will then connect the battery pack's (red) power line to the Pico's VSYS (5 V) pin, then the (black) ground line to a GND pin (**Figure 1**). We're using standard jumper wires for this basic test, but there will be a slight voltage drop depending on the thickness and length. Ideally, thicker 18AWG wires should be used.

With the wires connected to the correct pins, Pico should power up. An easy way to check is to first install a simple script to blink its on-board LED. With the Pico loaded with MicroPython and connected to a computer, save the following code as **main.py** (so it autoruns when powered up). →

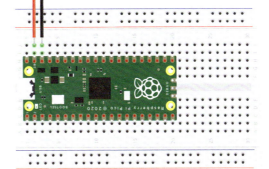

Figure 1
Wiring diagram for connecting a battery pack directly to Pico's VSYS and GND pins – not ideal, but it does work

Below
A small selection of regulator boards. From left to right, a Pololu S7V8F5, Adafruit PowerBoost 500C, and DiP-Pi PIoT for Raspberry Pi Pico

VOLTAGE REGULATORS

While you could always build your own DIY regulator circuit, using resistors in series as a 'voltage divider', it's more convenient to buy an off-the-shelf voltage regulator board. There's a wide range available, suitable for almost every need.

A step-down (or buck) converter will reduce an input voltage to produce a lower output voltage; for instance, 5 V to 3.3 V. Conversely, a step-up (or boost) converter will increase an input voltage to produce a higher output voltage. Due to Ohm's law, this affects the relative current level of the input and output; in a step-up converter, the output current level will be reduced. Also, as no converter is 100% efficient, you should expect to lose some of the overall power level in the process.

There are versatile boards, such as the Pololu S7V8F5 regulator that we used, that can step an input voltage up or down to maintain a set output voltage. Lower-cost regulators are available, but a key factor to consider is the tolerance of their output voltage – if it's 10% on a 5 V output, it could creep up to 5.5 V, which is above the maximum 5.25 V input voltage for a Raspberry Pi SBC (Raspberry Pi Pico is more tolerant).

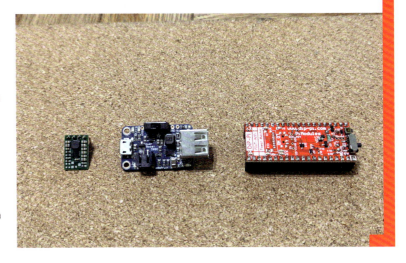

```
from machine import Pin
from utime import sleep

led = Pin('LED', Pin.OUT)

while True:
    led.toggle()
    sleep(1)
```

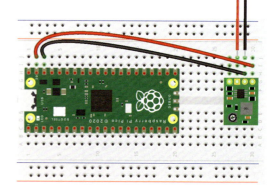

Figure 2 🔗
Wiring diagram of a battery pack connected to Pico's pins via a voltage regulator

You should then see the Pico's green LED blinking when you connect your battery power source.

Since Raspberry Pi Pico can work from a supply as small as 1.8 V, there should be no issues getting it working from even fairly low-capacity batteries. A board such as a Raspberry Pi single-board computer or Arduino is a different prospect, as it has a narrower voltage tolerance range – typically from 4.75 to 5.25 V. Any lower and it may well not stay powered for long, if at all, in particular under a heavy workload or with any peripherals connected.

THE BUCK CONVERTER STOPS HERE

To obtain a more reliable battery power supply to the device, it's advisable to use some sort of voltage regulator. A buck converter is able to take a higher DC voltage and convert it to a lower one while increasing the current. This is typically used with a higher-voltage battery source than needed, so long as it's within the converter's input range. This is a good option when powering the board and other parts of a project separately from a single higher-voltage source.

Since our 3 × AA battery pack is supplying less than 5 V, we opted to use a Pololu S7V8F5 5 V Step-Up/Step-Down Voltage Regulator (**hsmag.cc/PololuS7V8F5**). As its name implies, this tiny board can also 'step up' (i.e. increase) a lower voltage to 5 V, with a slight loss of current – you don't get anything for free. Based around the TI TPS63060 regulator, the S7V8F5 can accept voltage in the range 2.7 V to 11.8 V, making it very useful for supplying a regular 5 V output to boards and other components in a project.

After soldering some header pins to the S7V8F5 and plugging it into a breadboard, we connected the

A **HIGHER POWER**

If you need to supply a more demanding board such as a Raspberry Pi 3B/B+, 4 or 5, you'll need a 5 V step-down converter and a higher-powered source in order to maintain the current necessary to keep it running. For instance, a Raspberry Pi 4 requires at least 700 mA just to boot up, but typically more (around 1.25 A) when running any workload with peripherals attached.

The situation with the Raspberry Pi 5 is complicated by its use of PD (Power Delivery) to deliver more current to its USB ports. It normally requires a 27 W mains power supply to be able to run at full pelt. You're unlikely to need this for a portable, battery-powered project, but then you might be be better off instead using a Raspberry Pi Zero 2 W, which has a typical bare-board active current consumption of 350 mA.

battery pack wires to the appropriate inner pins (VIN and GND) and then the outer 5 V pin and inner GND (again) via jumper wires to VSYS and GND pins on our Raspberry Pi Pico (**Figure 2**). You can ignore the outer SHDN (shutdown) pin, which is only used to put the S7V8F5 into low-power mode.

As you'd expect, the fairly voltage-tolerant Pico powered up without any problems. Now to try it with a Raspberry Pi Zero. With a steady 5 V supply from the Pololu S7V8F5, it powered up successfully. With the generally low power draw of the Raspberry Pi Zero, you can get away with a relatively low current (typically 80 to 120 mA).

With more demanding Raspberry Pi models, however, the maximum level of current supplied may not be enough. In which case, you can simply connect a beefier battery pack to the S7V8F5 – up to 11.8 V, although you still need to beware of possible LC (inductor-capacitor) voltage spikes caused by an initial rush of current when first connecting it. If the supply is 9 V or more, Pololu advises keeping connections short and/or connecting a 33 µF or larger electrolytic capacitor close to the regulator between VIN and GND. See **pololu.com/docs/0J16** for more information on LC voltage spikes.

LIPO SUCTION

Next, we thought we'd try powering our boards using a LiPo (lithium polymer) battery cell, which offers a compact, more convenient alternative to regular batteries. Now, most LiPo cells output 3.7 V, so unless you're using a Pico or another board that isn't too fussy about the input voltage, you'll need a buck converter to step it up to 5 V.

There are two key factors to consider when choosing a LiPo battery: the capacity, in mAh (milliamp hours), which shows how much power can be stored when it's fully charged; and the 'C' rating, which signifies the maximum continuous output that the battery can maintain. To find the maximum discharge current level, multiply the capacity by the C rating: for instance, a 2000 mAh battery with a 1C rating can output a maximum 2 A for one hour, although this current level may be reduced if using a buck converter to boost the voltage.

> You will need a special charger **module to recharge your LiPo cell**

Naturally, the bigger the power requirement of your device, the quicker the battery will drain. Another thing to note is that as it drains, the output voltage will typically drop – another good reason to use a voltage regulator.

Most modern LiPo cells have a built-in protection circuit to prevent overcharge, overdischarge, and overcurrent, as well as offering short circuit and over-temperature protection. Make sure yours does. You then won't have to worry about it. You will need a special charger module to recharge your LiPo cell, however. →

BATTERY TYPES

Let's take a look at some of the most common battery technology types:

Alkaline: Standard batteries that are designed to be single-use, so don't try charging them – it can be dangerous!

NiMH: Rechargeable batteries that come in the same standard sizes as alkaline ones. The downside is that they typically have a slightly lower voltage output: 1.2 V compared to 1.5 V for AA batteries.

LiPo: Lithium polymer cells that can be recharged. Their slimline nature makes them ideal for tight spaces. They typically output 3.7 V, so you'll need a step-up regulator to get 5 V. Most LiPo cells have built-in circuitry to prevent overcharging and to shut off the output when the voltage drops too low, to avoid damage to the battery.

Li-ion: Lithium ion cells that are rechargeable and come in a range of shapes and sizes, including coin cells. Cylindrical sizes include the popular 18650, along with AA and AAA. They have a higher power density than LiPo batteries and don't suffer from the 'memory effect' that makes cells harder to charge over time. There are versatile boards, such as the Pololu S7V8F5 regulator we used, that can step an input voltage up or down to maintain a set output voltage. Lower-cost regulators are available, but a key factor to consider is the tolerance of their output voltage – if it's 10% on a 5 V output, it could creep up to 5.5 V, which is above the maximum 5.25 V input voltage for a Raspberry Pi SBC.

Above ◆
Clockwise from top left: LiPo cell, 18650 Li-ion pack, alkaline batteries, NiMH rechargeables

Above ◈
LiPo batteries are a good, slimline option for portable power. Here, we've connected one directly to a Pico

When it comes to connecting it up, most standard LiPo cells have a JST 2-pin or Deans connector (the latter type mainly used for RC vehicles and drones) – you can either plug male jumper wire ends into this for your connection or chop the connector off and tin the wire ends, perhaps for use in a screw terminal block on a buck converter.

With a fully charged 1000 mAh LiPo battery connected via a Pololu S7V8F5 regulator to a Pico W, we measured the output voltage at a stable 5.11 V, while the current level was around 60 mA with the on-board LED blinking. At this discharge level, the battery could in theory last for up to 16.5 hours, but in reality it will be less due to inefficiencies and the battery automatically shutting off power when it reaches a low level.

Other devices may draw more (or less) power. A Raspberry Pi Zero W typically draws between 80 and 120 mA, depending on whether it's run headless or not and whether Wi-Fi is used, so a 1000 mAh battery should in theory be able to power it for up to 12.5 hours (1000/80), although it's likely to be shorter due to inefficiencies.

UNINTERRUPTED POWER SUPPLY

As we've seen, batteries can be used to power a portable project for a limited period. While this may be fine for many projects, such as robots or a handheld games console, it's not suitable for those that need to be kept powered up for long periods, or continuously. In that case, your battery power can be used as a backup for a mains supply – or even another portable power source – to provide an uninterrupted power supply (UPS). This is ideal for projects that need to be kept online continuously to log sensor data, such as a weather station.

There is typically no power regulation or fuse protection... to protect the board from overvoltage or current spikes which may damage or even destroy it

Even in the event of a power outage, they will stay powered up.

For this, you'll need a UPS board that can instantly switch the supply from mains to battery backup when required. Various such boards are available, including the Adafruit PowerBoost 1000C which doubles as a voltage regulator.

We tried out a Waveshare UPS HAT for Raspberry Pi Zero W/WH (**hsmag.cc/WSZeroUPS**), which is mounted on the underside of the latter, connecting with springy pogo pins – thus leaving the GPIO pins free. It also comes with a 1000 mAh LiPo battery which can be charged when the HAT is plugged into the mains. It supplies a regular 5 V supply to the Raspberry Pi Zero, on which you can run a Python script to show useful battery info such as the real-time load voltage, current (negative if battery power is being used), power, and percentage.

Other UPS boards may offer even more detailed battery data and features. A good option for Pico, the DiP-Pi range (**hsmag.cc/DiP-Pi**) offers multiple power inputs from 6 V to 18 V, battery charging and monitoring, and various extra features. For Raspberry

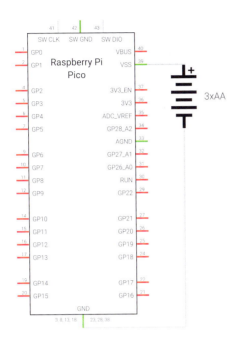

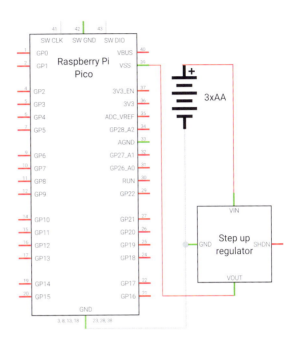

Above ◈
Schematic (left) for
connecting a battery
pack directly to
Pico, and schematic
(right) for connecting
it via a voltage
regulator board

Pi SBCs, the PiJuice HAT (**hsmag.cc/PiJuice**) offers advanced power management features including a 'watchdog' timer that monitors a software 'heartbeat' and, if it's not heard for a certain period, automatically resets the Raspberry Pi.

A UPS board is also ideal for connecting alternative power sources such as solar panels, enabling you to charge a connected battery to maintain a continuous supply for remote projects away from a mains power source, such as a weather station or wildlife camera.

PORTABLE POWER OPTIONS

In conclusion, yes, you can power a Raspberry Pi or other board directly from batteries to the GPIO pins and it should work, providing suitable voltage and adequate current are supplied. However, there's nothing to protect against issues such as voltage spikes and dropouts which may damage the device or even the battery (if it lacks on-board circuit protection). For an inexpensive microcontroller, you may not be too bothered about that, and many of them can take a wide range of input voltages – from 1.8 to 5.5 V in the case of a Pico.

If you don't want to run the risk, you're much better off using some form of power regulator to ensure a constant voltage and offer protection against spikes and other issues. Some of these boards offer advanced features, too, enabling you to monitor the battery level and output, and in some cases even run custom scripts when certain events are triggered.

In addition, a UPS board will ensure an uninterrupted power supply backed up by the battery and can be used with renewable power sources such as solar panels so that you can maintain a power supply to critical projects that need to keep running continuously – for example, to gather sensor data over a long period of time. ◻

USEFUL RESOURCES

- **Raspberry Pi power requirements**: hsmag.cc/RPiPower
- **Raspberry Pi Pico power**: hsmag.cc/PicoPower
- **Battery University**: batteryuniversity.com
- **Circuit Journal Arduino power guide**: hsmag.cc/ArduinoPower

Parts bin wings

We shamelessly adapt an existing design with the parts we have on hand

Ben Everard

Ben spends more time making games controllers than he does playing games. It's a strange hobby, but he enjoys it.

We saw an excellent tutorial from Adafruit on making some animatronic wings: **hsmag.cc/AdafruitWings**. They looked awesome! The only problem was, we didn't have the parts needed to make them. Obviously, we could have ordered all the parts we needed from those good folks in New York, but we did have a lot of similar parts. We wanted to know, could we reproduce the wings using just what we had hanging around the workshop left over from previous projects?

Re-using stuff you have is a great practice to get into. It's cheaper, better for the environment, and means you can make stuff now rather than waiting for a delivery (and this can be key to finishing a project rather than ending up with a pile of half-finished projects).

It does sometimes mean suboptimal results – we'll look at some of the areas where our wings aren't as good as the originals. However, as the saying goes, done is better than perfect. We set ourselves the challenge of building this without buying anything new. Obviously, you won't have the same bits and pieces lying around, but we'd encourage you to use what you have and improvise where necessary.

A SOLID BASE
We won't go through the whole build as the base is documented well on the Adafruit website. We'll focus on the parts where we've differed from Adafruit.

At the core of the project is a pair of servos, and we had some the exact same size, so we were off to a good start. The other parts we had that matched exactly what was used are a pair of 608-ZZ bearings (we used them to make a filament spool holder and had a few left over) and a lithium battery (that has been through a few projects already, and regular readers will recognise from an extremely janky pair of headphones).

That's not a bad start. However, we didn't have a Feather M4 Express, a Servo FeatherWing, a FeatherWing Doubler, a PowerBoost Charger, or any of the nuts and bolts used.

The first thing we had to work out was if we could mount the servos, as the servo arms designed by the Adafruit team screwed into the bearings. After a bit of fiddling, we worked out that rather than attach these arms to the bearings with a machine screw, we could simply add a protrusion to the arm. This had the downside of forcing us to print the arm in a less favourable orientation, which both meant additional support material and layer lines in a weaker position. However, in practice, they printed OK and seemed strong enough.

With the servos in place and working, it was time for our next hardware flip – changing the Feather M4 for a Raspberry Pi Pico. The short version of this story is that it just worked. The longer version is that it's actually really impressive that you can switch controllers like this. While both the Feather M4 and the Raspberry Pi Pico are built on Arm CPU cores,

they're built on different Arm cores with completely different things around these Arm cores. The fact that it works so easily on different microcontrollers is thanks to the work of the CircuitPython developers.

We swapped the PowerBoost board for a Pimoroni LiPo Amigo board. This was swapping a similar component for a similar component. The biggest impact of the microcontroller and power board swap is that they no longer fitted on the mounting plate. At this point, we had two options: design a new mounting place, or hack it together. We opted for the latter of these two options. Our workshop is currently devoid of any machine screws below M4, so we couldn't have built it without ordering more hardware anyway, and that's against our self-imposed rules.

We now have a bunch of parts and no nuts and bolts to join them. We do, however, have a tube of superglue (also known as Krazy Glue or cyanoacrylate). Sometimes the easiest solution is perfectly acceptable. Yes, superglue probably isn't as secure as bolting everything together, but then the wings are by far the weakest part, so it's really not the end of the world if the base is slightly weaker than it could be. Gluing them on does mean that the cables are a bit more exposed than they could be, so we used plenty of cable ties to tie the cables down and keep things as tidy as possible.

That's just about all there is to it. We didn't add the potentiometer, because we didn't really need the features it offered.

It would have been slightly stronger and noticeably more tidy if we had built it as it was designed, but then it's strong enough and tidy enough as it is, and with many builds, speed is of the essence. The quicker you can complete something, the more likely you are to see it through to completion.

Waiting for parts to arrive is a recipe for distraction and waning enthusiasm. At least, that's our experience (and our excuse). ▢

Above ↖
We're not completely sure if this Raspberry Pi Pico will come off the glue without breaking, but given that, at £4, it's about the same price as a delivery of bolts to screw it on, we're willing to take the risk

Below Left ↙
We were late arrivals to the world of cable ties, but now keep some in different sizes on hand because they're incredibly useful

Raspberry Cash

Build a working cash register with a Raspberry Pi

Dr Andrew Lewis

Dr Andrew Lewis is a specialist fabricator and maker, and is the owner of the Andrew Lewis Workshop.

Right ◆
Yes, you could always just wear a trader's money belt and carry a calculator if you need to, but a battery-powered cash register with a working drawer looks much nicer. Being able to take cash and generate receipts without a power connection can save hundreds of pounds in cash if you're at a trade show where the organisers charge for power connections

QUICK TIP

If you want to delete the last item scanned, use the **BACKSPACE** key. If you have text in the register's buffer, it will delete the last character typed. If the buffer is empty, it will delete the last line item added to the current sale.

The world of the future is a largely cashless society, but the world of today still uses metal tokens and paper promises to facilitate transactions. In an urban store you'll probably find an electronic point of sale (EPOS) system that handles the shopping experience, but when you're on a market stall away from regular power, the conventional EPOS experience isn't as easy to implement. In this article, you'll see how to create a working cash register with a secure money drawer, barcode scanner, and receipt printer. You'll get it all running from a Raspberry Pi and DeWalt battery, and you'll also have a built-in daily transaction log, and the option to use any wireless device as an extra till display.

CALCULATOR WITH CONTROL ISSUES

Cash register and EPOS system are really just catch-all terms for a fancy calculator, and it's worth defining exactly what we're going to be building here, and why. A typical modern cash register sits on the counter next to the cashier and it allows the cashier to calculate the total value of the items being purchased. In the old days, this was a purely mechanical beast that required the cashier to know the price of each item and enter it correctly. As technology advanced, barcodes were used to give each item a unique code that could be read by the cash register and checked against a database of prices, so that the cashier no longer needed to enter the items manually into the register unless the barcode was damaged. In modern times, the cashier

REGISTER INPUT

The cash register application gets input from the barcode scanner and keypad by monitoring system-wide key presses using pynput. Key presses from the touchscreen web interface are generated by passing an API request from the web page back into the main app, which generates the appropriate key codes using pynput. It's worth noting that the generated key code may vary between OS and the type of device that generates them. For example, the decimal point may generate different codes depending on whether it's being sent from the key on the main keyboard, the numeric keypad, or an external device. You may need to adjust the Python code slightly to suit your particular input devices.

The barcode scanner used in this project appears as a standard input device, and to all intents and purposes acts like a keyboard. When the user scans a barcode, the code is translated into a string of characters and 'typed' into the computer. To avoid the need for a database of scannable items, the barcode is encoded as a simple string with a ':' symbol as the delimiter. The first part of the string is an item description, and the second is the price of the item. So a barcode with the string 'snacks:5.4' would generate a sale item with the title 'snacks' for £5.40 (or whatever currency you're working in). The type of barcode isn't actually very important: as long as it can encode the text you want to enter correctly, the scanner will read and decode it as plain text.

that displays the current transaction in a browser. The browser launches in kiosk mode when the computer starts. This might seem a strange way to implement the display, given that there are several GUI tools (like Pygame, PyQt, or Tkinter) available for Python, but it's actually a very flexible way to implement the interface. The web browser can be easily customised with HTML and CSS to alter its appearance without needing to change the main Python code, and additional interface functionality can be added using JavaScript on the client side. You can also connect multiple web browsers to the same cash till and use them as remote displays or remote terminals for the register.

For this project, you will be connecting the Raspberry Pi to a basic receipt printer, which requires between 5 and 9 volts to power it. You may have already spotted the immediate problem that the Raspberry Pi GPIO header uses 3.3 V and isn't 5 V tolerant, although this isn't actually a huge issue. The printer only needs bidirectional communication to report the status of the paper tray (which isn't really necessary for everyday use) and the 3.3 V GPIO voltage of the Raspberry Pi is high enough to trigger a high input on the printer, so it's possible to connect the printer directly to the Raspberry Pi without a level shifter, →

is sometimes entirely missing from the equation and the shop makes us do all of the work ourselves, paying for items by putting money into a slot or waving a debit card in front of a scanner. What you will be creating is something like the last generation of cashier-driven register. Your cash register will have a touchscreen interface, a barcode scanner, a remote keypad, a receipt printer, and the facility to keep daily transaction lists that can be read at a later date. This register won't accept debit or credit cards by default, but it should be possible to add that functionality if you want it. With a little extra work and a third-party API like Square, you could generate transactions that accept card payments from a contactless reader.

The cash register application is written in Python, and uses a Flask web server to generate a web page provided that you don't connect to the RX pin (GPIO 15) on the Raspberry Pi.

You can normally force a receipt printer to produce a self-test by powering it on with the button pressed. Although it varies from brand to brand, you'll normally see information about the firmware, character sets, and serial connection settings for the printer. Most serial ports will be defaulted to 9600 or 19,200, with the usual 8 bits, no parity, and 1 stop bit.

Thermal printers use a control language called ESC/POS, using escape codes to generate character effects. It's an old technology – and unlike a modern desktop printer, a thermal receipt printer has very limited capabilities. Printing images is quite complicated to achieve, and printed text relies on a built-in font with escape codes used to apply simple modifiers like reverse-printing and double-size characters. There's also a built-in facility to generate barcodes in most receipt printers. Some units also have a special connection that can be used to trigger the cash drawer when a receipt is printed.

While it's possible to set up a serial connection to the printer and send the escape codes directly, it's much easier to use Adafruit's Thermal Printer library and Blinka. Blinka allows you to use CircuitPython APIs in regular Python, and can be installed from **circuitpython.org/blinka** or using the pip installer.

QUICK TIP

Thermal printers are fussy and easy to confuse. If you're getting garbage printed unexpectedly, try power-cycling the printer to reset it.

Below ◆
Thermal receipt printers are very simple and communicate using an RS-232 or TTL port, although some of the more modern units have a USB or Bluetooth connection

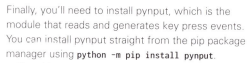

DRAWER **TRIGGERS**

Cash drawers are normally available in 12 V or 24 V varieties. The 24 V version is probably more common on modern EPOS systems, while the 12 V versions are usually found on older-style cash registers. Some new cash drawers are smart and have a processor built-in with a USB trigger, but in general, most 'normal' cash drawers just have an RJ11 socket that connects to the solenoid and to a microswitch that triggers when the drawer is open. On these cash drawers, you can easily map the pinout of the wires by measuring the resistance between the pins on the RJ11 socket. The resistance of the solenoid will be easy to detect with a multimeter, and the continuity between the pins connected to the microswitch should also be easy to detect. Opening the drawer is as simple as applying the correct voltage to the solenoid for a short amount of time (typically less than 250 ms). On some of the more complicated drawer systems, there are two independent cash drawers with separate solenoids that share a centre tap. This is sometimes referred to as 'Epson wiring' with pins 2 and 5 connected to the outside (negative) legs of the solenoids, while pin 4 is the shared (positive) centre tap. There are several other systems in use, but these are probably the most common in the modern day.

Finally, you'll need to install pynput, which is the module that reads and generates key press events. You can install pynput straight from the pip package manager using `python -m pip install pynput`.

One thing to watch out for with thermal printers is that they are relatively slow and don't really communicate bidirectionally with the device they're connected to. This means that there's no signal to say when the printer is ready to print, or when the printer is already printing. It's quite possible to send data to the printer so quickly that the internal buffer becomes overwhelmed and it starts missing characters. You need to make sure that the program you're using won't send the data to the printer faster than it can output onto the till roll.

Begin making your own cash register by installing the latest version of Raspberry Pi OS onto a Raspberry Pi 3, or later model, and making sure that all of the packages are up to date. Once you're ready, go into the Preferences menu on the desktop and bring up the Raspberry Pi Configuration application. Check that the serial port is enabled, and serial console is disabled, accept any changes, and reboot the machine.

Next, you'll need to install some Python libraries that the cash register application depends on to get your code working. You'll need the Blinka library and the Adafruit Thermal Printer library, which you can install by following the instructions at **hsmag.cc/AdafruitReceiptPrinter**. Next, you'll need to install the Flask framework, which you will be using to create a Python-based web server. Follow the instructions found at **hsmag.cc/Flask** to do this. Then, install the cash register files into a folder called **till** on the desktop (or in the location of your choice). The main application in the folder is **keys.py**.

Next, you need to wire up all of your electronic components to the Raspberry Pi's GPIO pins. You'll be drawing power from an 18V battery, and dropping to 5V and 12V using buck converters. The 12V buck converter is used to open the cash drawer, and can →

KIOSK **MODE**

You'll probably want to automatically run the **keys.py** file and start a kiosk mode web browser when the computer starts up. To do this, you need to install some packages that will help you control the window manager, and edit the autostart file in **/home/user/.config/lxsession/LXDE-pi**.

Begin by installing the packages:

```
sudo apt-get install x11-xserver-utils unclutter
```
Edit the autostart file using nano:
```
sudo nano /home/pi/.config/lxsession/LXDE-pi/autostart
```
Add the following lines to the autostart file:
```
@python /home/user/Desktop/till/keys.py
@chromium-browser --kiosk --incognito --disable-pinch --overscroll-history-navigation=0 http://127.0.0.1:5000
@xset s noblank
@xset s off
@xset -dpms
@unclutter -idle 0.1 -roo
```

Save the file and exit nano. This autostart configuration will cause the Raspberry Pi to boot into a kiosk mode, disable the screensaver, and hide the mouse. It'll also mean your desktop will be completely blank. If you want to be able to access the desktop, you'll need to edit the autostart again to include the following lines at the top of the file:

```
@lxpanel --profile LXDE-pi
```

QUICK TIP

The more powerful the Raspberry Pi, the more battery power it will need to run. A Raspberry Pi 3 will be fine for this project.

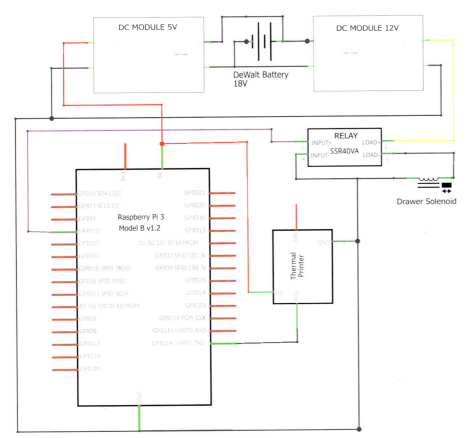

Left
You can see from this diagram that the wiring for the cash register is not complicated, but there are a couple of things that you should watch out for. Firstly, this diagram doesn't show the USB or HDMI connections for a touchscreen or other peripherals. When you are mounting your Raspberry Pi inside your case, be sure to allow enough space for cables to plug in. You should also apply the same thinking to make sure that you can get batteries in and out easily. Remember that any digital cable is susceptible to interference, so keep the cables to your external components as short as possible

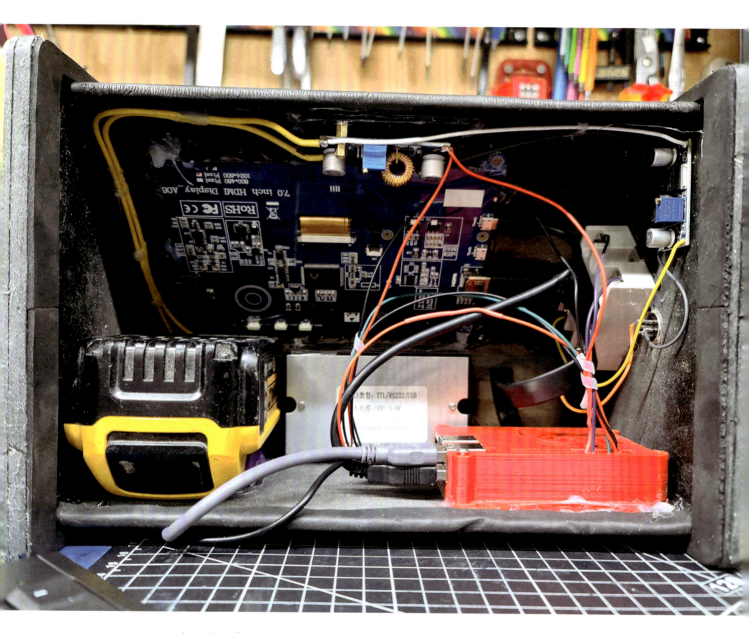

be omitted if you have a 24 V cash drawer. If you
have a 12 V drawer and want to risk omitting the buck
converter anyway, the risk and responsibility are your
own to consider. It may work fine, or it may burn out
the drawer solenoid after a while.

The 12 V power is controlled by a solid-state
relay connected to GPIO 17, and should only trigger
for a few milliseconds at a time, which would
probably prevent the solenoid from damage, but
that isn't guaranteed safe to use. The other 5 V
buck converter is used to power the Raspberry
Pi and the thermal printer. It's not unusual for

thermal printers to print out their input voltage
during a self-test. This isn't very accurate, so don't
be too worried if the reported voltage is slightly
under 5 volts. The thermal printer is also a power
hog, and can draw up to 10 W while printing. The
Raspberry Pi itself (and USB peripherals) will only
be drawing a couple of watts, but if you are using
a touchscreen display, it can add another 4 or 5
watts to overall consumption depending on how it's
configured. If you're planning on a long workday,
consider taking along extra batteries just in case you
start running short.

Below ◈
This project is likely to get bounced around a bit when you're setting up and disassembling your store. Make sure that your case is sturdy enough to deal with that by gluing your joints well

Left ◈
There's plenty of spare pins for expansion of the cash register, so adding NFC readers, servos, or custom button boards is absolutely possible

There are a few extra things that you could consider adding to this project if they suit your needs, including adding an NFC reader to read staff cards so that you can record who took a sale, or adding extra 'hidden' web pages to make downloading daily transaction records easier. You could also add a second battery connector to double the battery life and make it possible to swap batteries without restarting the till. You could also implement a local database of items with prices, so you can set the prices at the till rather than on the barcode itself. ◻

Pico keyboard and gamepad

Two ways of using a Raspberry Pi Pico to control a computer

Ben Everard

Ben spends more time making games controllers than he does playing games. It's a strange hobby, but he enjoys it.

Above ◈
The easiest way of building a controller like this is to solder some buttons onto protoboard

The first thing that you usually learn when doing electronics with a microcontroller is blink an LED. The second is connecting a button. In this article, we're going to take this second skill and use it to build a peripheral for a computer. There are surprisingly large stacks of software between the switch on a keyboard and the program you use it to control. Fortunately, though, we don't have to concern ourselves with them. We can grab some pre-written firmware, adjust a bit of configuration, and then have everything 'just work'.

Buttons come in many different shapes and sizes, but they mostly do the same job – bring two contacts together when you press them. For our microcontroller to read the state of the button, we just have to connect it between a GPIO pin and ground.

You can hook up GPIO pins like this on many microcontrollers, but there are a few features that make some work better than others. The first is internal pull-ups. To read the position of a button, you need to connect one part of it to ground, but you also need to add a resistor between the GPIO pin and a positive voltage. Since this is a very common thing to

GETTING TO **GP2040-CE WEBCONFIG**

You configure GP2040-CE using the web-based config tool. To get there in the first place, you have to hold down button S2 while you plug in your controller. However, if you don't have a button S2, you can flash the force-webconfig version as we have done. If you want to get back to the web config, you can just flash it again.

Left ◈
Once all the components are in place, you can solder wires between them to connect them up as needed

do with microcontrollers, many have built-in resistors, and these are known as internal pull-ups.

The second feature that will prove useful to us is USB device support. Most microcontrollers can plug in via USB, and be programmed that way, but depending on the implementation of the USB software and hardware on the microcontroller, some can be programmed to be keyboards, mice, and gamepads.

We won't go into these in too much detail other than to say that the Raspberry Pi Pico is an excellent choice for both of these reasons, and it also has another little trick up its sleeve – an abundance of ground pins. These mean that you can wire up buttons individually without having to worry about the logistics of connecting all the buttons to ground.

ON / OFF

You can connect the switches up using whatever method you feel comfortable with. You can use a breadboard, you can solder wires between them, you can protoboard them – it really doesn't matter. Just make sure that each switch is connected between a GPIO pin and ground.

We're going to make a games controller first, so you're going to need at least six buttons (four for direction and two action buttons). Once you've wired them up, it's time to add the software.

We're going to use the GP2040-CE firmware. This takes an RP2040 board and turns it into a high-performance gamepad. It takes care of all the tricky things like communicating with the computer and checking for button presses. All you have to do is tell it which GPIO pin relates to which button. You can get the firmware from **hsmag.cc/GP2040**.

ENTER **THE MATRIX**

You might have noticed that Raspberry Pi Pico has 26 GPIOs you can use to attach buttons to, but that would only satisfy the most militant of minimalist keyboard users. There are some microcontrollers with more GPIOs, but they aren't particularly common (at least in the hobbyist community), and dev boards breaking out all these connections can be expensive. How, then, are we able to make DIY keyboards?

The answer is a matrix. Rather than connecting up each button to its own GPIO pin, we create a grid of wires horizontally and vertically. We can then place switches joining a horizontal and vertical wire at the crossing points.

We can then use the GPIO pins on one axis as output, and the other as inputs. By providing power to just one row at a time, we can use the inputs on the columns to detect which buttons in that row are currently pressed.

We can scan through the rows quickly to detect presses at any place on the matrix with only a few GPIO pins. The only potential problem is that multiple button presses can create alternative paths through the matrix which can cause problems. The solution is to use diodes – a component that allows electricity to pass in only one direction. Using this combination of a matrix layout and diodes, you can connect up to 169 buttons to the 26 GPIO pins on Pico.

There, you'll see versions for lots of different controllers all built around the RP2040 microcontroller. The one we want is the **force_webconfig.uf2** file. Download this and flash it to your Raspberry Pi Pico.

Once it has been flashed, you should be able to access the web editor by pointing your browser to **http://192.168.7.1**. →

On this website, you can go to the 'Pin Mappings'
section and select which GPIO pins you want to be
responsible for what actions. Once you've entered
everything and clicked Save, you can reboot your
Pico as a gamepad. It's a regular USB gamepad, so
you can use it for your favourite games in just the
same way you'd use any other gamepad. Want more
features? No problem, you can add them. As well as
adding buttons, you can add analogue inputs such as
joysticks and feedback from LEDs and buzzers.

TINY KEYBOARD

GP2040-CE is great for making games controllers,
but what if you want to control a game that takes
input from regular old keyboard presses? We can
do that too using the exact same circuit as with our
gamepad. There's a sizeable community of DIY
keyboard makers who have created some
excellent and hackable keyboard firmware. The one
that we'll be using is called KMK, and it is
based on CircuitPython.

To use it, you first need to install CircuitPython
on your Pico. To do this, head to **circuitpython.org**
and download the latest version for Raspberry Pi
Pico, then flash it to your device. You should find
that when you plug in your Pico now, a USB drive
called 'CircuitPython' appears.

Now, you'll need the KMK firmware from
hsmag.cc/getkmk. Unzip the file and copy both
the **kmk** directory and the **boot.py** file to the
CircuitPython drive. To turn this into a working
keyboard, we need to add a **code.py** file that
describes the particular keyboard.

JOYSTICK

Joysticks come in two forms: analogue and digital.
You can get either type in modules that are easy to
wire up to a microcontroller. Analogue joysticks are
basically two potentiometers, set up so each one
reads a separate axis: one for X and one for Y. Each
potentiometer has three connections: one for 3.3 V,
one for ground, and one output. You can wire them
up to an ADC input on Pico and read them using the
GP2040-CE firmware.

Digital joysticks are basically a stick with four buttons,
with a different button being pressed depending on how
you move the stick – it's basically a D-pad arrangement
rotated in on itself with a stick to do the pressing. In
these joysticks, there is usually a single common ground
connection, and then a separate GPIO connection for
each button. You can wire them and connect them to
either gamepad or keyboard firmware.

The code for our keyboard is:

```
import board
from kmk.kmk_keyboard import KMKKeyboard
from kmk.scanners.keypad import KeysScanner
from kmk.keys import KC


# GPIO to key mapping - each line is a new row.
_KEY_CFG = [
    board.GP2,  board.GP3,  board.GP4, board.GP5,
    board.GP9,  board.GP13, board.GP17, board.GP21
]

keyboard.keymap = [
    [
        KC.UP, KC.DOWN, KC.RIGHT, KC.LEFT,
        KC.A, KC.B, KC.C, KC.D,

    ],
]

# Keyboard implementation class
```

Right
Thanks to the number of ground connections on Pico, we can solder on all the buttons without complicated wiring

Below
The joystick is really just four buttons that get triggered when the stick is moved

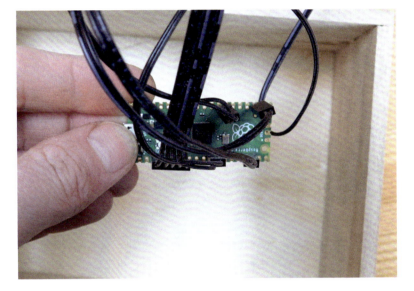

```
class MyKeyboard(KMKKeyboard):
    def __init__(self):
        # create and register the scanner
        self.matrix = KeysScanner(
            # require argument:
            pins=_KEY_CFG,
            # optional arguments with defaults:
            value_when_pressed=False,
            pull=True,
            interval=0.02,  # Debounce time in
floating point seconds
            max_events=64
        )

keyboard = MyKeyboard()


if __name__ == '__main__':
    keyboard.go()
```

OTHER FIRMWARE

We've used KMK because it's based on CircuitPython, which we're familiar with, but there are other options. QMK is probably the most famous of the keyboard firmware, and there are a lot of things built on top of it, including complex and powerful systems to allow multiple key presses to be interpreted in different ways.

ZMK is another option that's built specifically for wireless keyboards, but unlike KMK and QMK, we've not tested it out fully.

The main part here is the two lists: `_KEY_CFG` and `keyboard.keymap`. These two match up to link a GPIO pin with a key press. You might have to adjust the GPIO pins depending on where your buttons are attached.

Save that **code.py** file to the CircuitPython drive and you should be able to start using your Pico as a very minimal keyboard.

You can use this controller as is, or expand it to have as many buttons as you can cram on. It can be as simple or complex as you like, but the important point is that it can be customised to exactly the way you want to control your computer. Happy gaming! □

Getting jiggy with it

Advance your workshop skills with custom jigs

Dr Andrew Lewis

Dr Andrew Lewis is a
specialist fabricator
and maker, and is the
owner of the Andrew
Lewis Workshop.

Looking through a hardware store
catalogue, you'll find all sorts of
commercially made jigs. Jointing
jigs, pocket jigs, and routing templates
are used by tradespeople to complete
common tasks because time is money,
and jigs typically save time. They can also reduce
the chance of making a mistake while you're
working with a tool, which can cut down on waste
materials. If you're a hobbyist used to working on
one-off projects or an independent maker working
on small batches, you might dismiss the idea of a jig
as 'something professionals use', and be happy to
push through the project relying on the old mantra
of 'measure twice, cut once'. This is sometimes a
mistake, because jigs are one of the most important
tools you have at your disposal in the workshop, and
in this article you'll find out why.

MEASURE NONE, CUT HUNDREDS

Creating and using jigs while you are working can be
one of the most satisfying and productive changes
that you can implement if you're making the transition
from amateur to professional maker. When you create
a jig, you are demonstrating that you're thinking about
the process of making, about improving that process,
and about ways to make your job easier, cheaper, or
more accurately than you can with existing methods.

To be perfectly clear, being a 'professional maker'
doesn't mean that you always need to use a jig.
The jig police are not real and they cannot hurt you.
You will not be shunned by the community for
failure to use a jig. There are times when it's
appropriate to use a jig, and times when it doesn't
really make a difference. The key is that you
understand that doing the extra work necessary to
create a jig can have benefits.

Left ⬉
If you feel like you're spending more time setting up your machines than creating actual products, it might be time to step back and think about designing some jigs. While a jig might take a little extra time to develop at the outset, it can save hours of work in the future. This simple jig holds six wooden coasters in place for sublimation printing, and significantly reduces the time and material cost required to create a batch of coasters

Those benefits may manifest themselves in terms of future time saved, by increasing the accuracy or finish quality of your project, or by adding a safety barrier to a task that might go wrong and ruin a unique or expensive part. As a hypothetical example, drilling a hole halfway through the bottom of a boat that's in the middle of the ocean is theoretically possible using just a drill and a bit of masking tape to mark the depth that you need to stop drilling. It's possible to do this, but it's obviously a terrible idea to do it this way. Using a short drill bit in a jig that stops the drill from progressing too far into the bottom of the boat is a much safer option. Cutting a curve through a piece of expensive plastic is theoretically possible with a jigsaw and a pencil mark. It's much easier and more accurate to use a router with a jig to control the path of the blade. In these situations, using a jig is obviously a good idea. →

Above ◆
This rotary jig and Z axis jig are good examples of how jigs can add functionality to a machine as well as just enhance existing processes. This is something that often gets overlooked when people talk about using jigs – they can improve your workflow by adding features without significantly increasing costs. The rotary jig allows you to laser-etch onto a cylindrical object, instead of flat surfaces. The Z axis jig lets you perform milling operations on a lathe, which can be useful if you have limited space or funds for a dedicated milling machine

However, in a situation where you're making a one-off project, an elaborate jig can be more expensive and time-consuming than the project needs. Let's ditch the hypothetical examples and take a look at some real-world examples of how jigs can be used to improve your workshop life.

Most makers are familiar with laser cutters, and how they can cut or engrave very fine details into a material. However, the bed of a laser cutter is often just a meshed surface, and while this is fine for cutting out shapes from a sheet of stock material, positioning items precisely on the bed of the machine can be very tricky. If you need to engrave or cut something in a specific place, you need to create some sort of jig to hold your work in place. Some laser cutters have a camera that lets you see the bed of the machine and adjust the position of designs in software so that they are engraved in the right place on the target object. You can also use tape to indicate the home corner of the laser cutter and calculate the precise positioning from there. Neither of these solutions are particularly good in practice.

Firstly, in the case of a camera overlay, if you are engraving several identical objects you will need to adjust your design in software to align with each of the objects. The camera viewing the bed is not necessarily 100 percent accurate when it transforms the absolute position of the object from the bed of the machine to the monitor, giving an unexpected margin of error. Taping the corner of the bed and measuring out from there is problematic because the observed position of the tape will change slightly depending on the position of your head relative to the object you're placing, and it takes time to make fiddly measurements and adjust objects so that they line up properly. In some cases 'it looks straight therefore it's straight' just isn't a good enough methodology.

In a case like this, generating a disposable or reusable jig is an excellent way to tackle the problem. Let's assume that you have a series of ceramic tiles that you want to engrave with a repeating pattern. You want these patterns to align accurately with the edges of the tile. Start by measuring your tile, and then grab a piece of thin MDF sheet. MDF sheet is cheap and easy to cut with the laser, and you're going to use it to create a temporary jig. Glue the sheet to the bed of your laser cutter using a few dots of hot glue at the corners. Don't go crazy with the glue; think of what you're doing as tack welding with a glue stick. You don't need to be accurate with this, but do make sure that the board is flat against the bed of the machine, and that the edges of the board roughly align with the X/Y axis of the machine.

Use your laser cutter's software to create a new file with a cutting path somewhere on the area of the MDF sheet. The cutting path should match the measured size of the tile you want to engrave. Lock this path in position, and add an 'out-dent' at one edge so that you can get your fingers in to insert and remove the tile from the space you are going to cut out. Save the file.

Home the laser cutter, then cut the path in the MDF and remove the tile-shaped piece of MDF from the middle of the jig. The remaining MDF is glued into place, and if your laser is focused properly, you

should now have a way of positioning the tile on the bed of the machine with less than 1 mm accuracy. It's cheap and easy to remove when the job is done. All that remains is to add your engraving pattern in your laser cutter software and to disable the cutting path. Save the file again, and engrave tiles to your heart's content. Once you've finished processing the batch, you can pop off the hot glue and remove the template.

In this example, we used a tile to demonstrate how to create a disposable jig for a tile – but you can use this technique to position any shape or size of object, engraving multiple items at once if you have enough space on your machine. If you are familiar with LightBurn software, you probably won't be too surprised to discover that its community resources (and also YouTube) have numerous links to premade jig templates for common products like dog tags, leather key rings, and pendants. The tutorials may implement the jig with a different method from the technique described here, but the templates can still be used in the same way if you wish. →

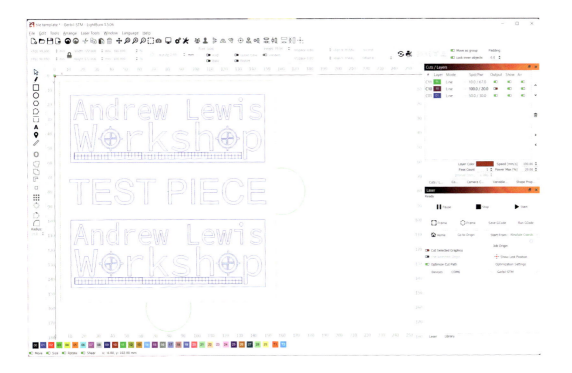

Left ◆
This LightBurn template has multiple layers that can be used to create a jig, and to engrave the surface of the tile. You can see that there are semi-circular cutouts along two of the edges to make it easier to replace the tile between engravings

QUICK TIP

If you can position the MDF against the edges of the machine in a repeatable way using fixed points, you can reuse the jig multiple times. If not, you'll need to throw it away and cut a new jig after it's been removed from the bed.

> **This sort of reusable jig does** several things at the same time

Above

The jig for this coaster was too thick, so the heated press couldn't effectively apply pressure across the whole surface. You'll notice that the black print on the right side of the coaster is pale and uneven because the contact with the transfer paper was poor

If you do have access to good laser software (like LightBurn) and you want to create a very accurate template, you can modify the kerf-offset settings to account for the thickness of the laser beam when cutting. By shifting the laser beam inside the shape you're cutting out, you can create a more exact-sized cutout. This might be useful if you are etching or cutting the solder mask on a PCB to expose traces, but for simple etching projects like key rings or business cards, this isn't necessary.

You might also use a jig to save materials or make it faster to create multiple items at once. Let's start out by thinking about sublimation printing some drinks mats or coasters. The basic process for sublimation printing is to print out the design in reverse using special ink and paper, and apply it to a specially prepared blank coaster with a heated press. This means lining up the coaster with the paper and holding the paper in place while the press is down. For a single coaster this is fine, but it gets fiddly and time-consuming when you're making multiple coasters. You need to print the designs, cut them to shape and tape them onto the coasters, then press them, let them cool, and remove all of the pieces of tape.

With a well-designed jig, you can hold multiple coasters in place at the same time, cutting down on waste. Beginning with another blank sheet of MDF, measure your coasters. The MDF should ideally be

about 1 mm thinner than your coasters are. Put the MDF into a laser cutter if you have access to one. You could use a fretsaw or another cutting tool if you don't have a laser cutter. You don't need to glue the sheet down this time. Depending on the size of your coasters, you'll probably have room for between four and six coasters on your sheet of MDF. Cut coaster-shaped holes in the sheet so that you have a few millimetres of border between each hole, and a few millimetres of space around the edge of the sheet. You want your jig to be smaller than the sheet of paper you'll be printing your designs on so that it's easy to fold the edges of the sublimation paper around the jig.

You're going to be using this MDF in a hot press, so make sure it's dry before you cut it by pressing it on both sides for a few minutes and then letting it rest. If it warps, it's still got some moisture in and needs more time to level out. Re-pressing may help if the MDF doesn't go flat after a while.

Print your designs onto the paper so that they line up with the holes in the MDF sheet and leave a print bleed around the edge. The print bleed is used to make sure any minor changes in the position of the coasters on the jig won't result in an unprinted edge. In other words, print the coaster designs slightly larger than the coasters actually are, and increase the border around the edge. This will mean everything will still look good if the coasters are slightly different sizes or the paper

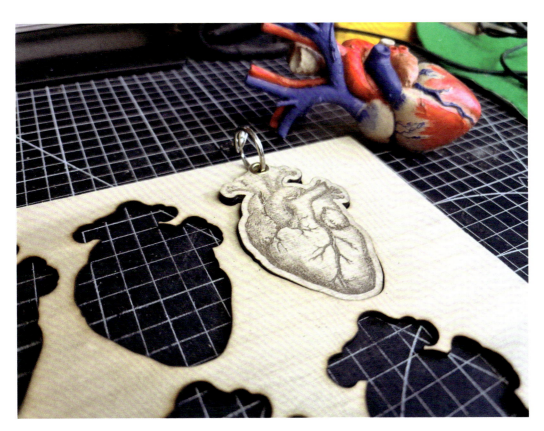

moves slightly during the pressing process. This sort of reusable jig does several things at the same time. It lets you treat several coasters as one object, which is more convenient to move around. It also means that you can eliminate the 'cutting out' step from your printing process, so instead of cutting out six individual designs, you simply print your six designs on a single sheet, align the sheet with the jig by eye, insert the blank coasters, and then fold the paper into place around the jig. Another benefit of this jig is that you can hold the paper for six coasters in place using two small pieces of tape, instead of the twelve pieces of tape that you need to produce the coasters individually. That saves money and reduces the amount of waste produced during the printing process.

While these examples show how you can create custom jigs yourself, there are plenty of pre-existing jigs out there that can be used to speed up your workshop and improve the quality of your work. Some jigs, like the shirt folding jig, are designed with the intention of regularising a high-level task (in this case, the name of the jig should give away the intended purpose) and make it easier to complete without experience. Other jigs serve a more fundamental purpose, like a drilling jig, which is created to improve the precision of drilling operations by holding the drill bit at 90 degrees to the surface being drilled. Tapping jigs and pocket hole jigs serve a similar purpose.

If you have access to a 3D printer, there's a huge range of jigs that you can create to make any number of workshop tasks less onerous. Even a cursory search will bring up printable jigs for common operations like tapping, indexing, tool-sharpening, circle-cutting, dowelling, edge-rounding, and shaping. Many other jigs can be downloaded for specialist tasks like guitar tuner drilling and aligning rocket fins. ▫

Above Left ◥
There are times when you're working with a laser cutter that you accidentally create a jig during the cutting process. The waste product from your project ends up with a perfect cutout shape of the parts, and this can be used as a jig or mask for finishing processes like painting or sublimation. You can see that waste cut from these heart key rings makes a perfect jig for holding the blanks while they are sanded and sprayed

PCTG

Move over PETG, there's a new glycol-modified terephthalate in town

Ben Everard

Ben's house is slowly being taken over by 3D printers. He plans to solve this by printing an extension, once he gets enough printers.

Below ◆
Excellent layer adhesion and toughness means PCTG works well for this tarp clip

If there was a prize for the most intimidating sounding 3D printer filament, poly cyclohexylenedimethylene terephthalate glycol-modified would probably win it, but in use, **PCTG** as it's more commonly known, is a straightforward plastic to print with.

We printed the filament at 265°C with the bed at 90°C. While we do print inside an enclosure, it's not heated. We just slightly modified a PETG slicer profile and it worked. This should be within the range of most home 3D printers.

Chemically, as well as symbolically, PCTG is very similar to PETG, and shares many of the more famous filament's properties, including durability and resistance to temperature and chemicals.

There are two big ways where PCTG shines in comparison to PETG: it's got a higher impact resistance, and it's got better layer adhesion. The marketing material tells us that it's glossier and more transparent, but honestly, we couldn't tell the difference. The YouTube channel CNC Kitchen has an excellent test on the mechanical properties at: **hsmag.cc/CNCKitchenPCTG**.

The downsides of PCTG compared with PETG are that it's available in fewer colours, and it's a bit harder to find. Historically, it has been more expensive, but the gap has closed significantly in recent years.

We tested this filament out with a range of different projects around this month's cover feature. It proved particularly good for tent pegs which can take a whack. On the other hand, the carabiners faired far better in PLA – despite the strength of PCTG, it's not as stiff as PLA, and that caused the carabiner to deform. In fact, PCTG isn't even as stiff as PETG, which isn't the stiffest plastic to begin with. This slight flexibility is part of the reason it can withstand larger impacts. Carbon fibre-reinforced PCTG is available for cases when you want PCTG's properties, but need it to be stiffer.

If you're not planning a trip to the great outdoors, there are still plenty of cases for PCTG. Most of the time this is going to come down to durability. The higher layer adhesion could also be great for objects that have to have some stress across the Z axis. It's not perfect, so this is still weaker than the X and Y axis, but it's better than most other easy-to-

print filaments. Potentially, this could save you from having to split an object up into multiple parts to print, and that could mean saving time and filament.

Perhaps the most significant downside – when compared to PETG – for us, is the lack of recycled options for PCTG. While recycling plastic is still an area fraught with misinformation, and doesn't always have the environmental impact we hope it could have, on the whole, recycled plastic is a significant ecological improvement over virgin plastic, especially oil-derived plastics like PETG and PCTG.

PCTG really deserves to be a more popular filament. We found it printed really well, and the combination of high-impact strength and high layer adhesion is a great combination for many mechanical parts. ◻

Above ◈
Despite advertised claims, we've been unable to persuade PHB filament to compost

Right ◈
The flexibility of PCTG makes it a poor choice for some applications, like this carabiner

PHA UPDATE

About a year and a half ago we tested out PHA filament which promised to be a truly compostable 3D-printable filament. Unlike PLA, which requires very specific conditions to break down – so specific that in reality it's not possible for most people to compost – PHA, it is claimed, will break down in normal home compost.

We placed a test print in a home wormery for over 18 months, and there's no noticeable degradation. While there may be conditions that it will break down in home compost, it's clearly not as simple as just putting it in with the rest of your compost and waiting.

Our wait for a truly compostable filament continues.

TUTORIAL ━━━━━━━━━━━━━━━━━━━━━━━━━━━━━━━━━━━━

Bring a 1960s spy camera back to life with 3D printing

Get pictures from a Minox microcamera by using 3D-printed technologies

Rob Miles

Rob has been playing with hardware and software since almost before there was hardware and software. You can find out more about his so-called life at **robmiles.com**.

Extend your photography and 3D printing skills by getting a classic camera back into action.

Minox 'spy' cameras are popular amongst camera collectors and those looking for a quirky way to take photographs – this despite their production having ended many years ago. The cameras are very well-made, and it is not hard to find working examples at reasonable prices. There are people who can provide you with film cassettes and develop them for you, but you can also 3D-print tools that let you produce your own pictures with this marvellous machine. In this article, we are going to look at the camera itself and find out what makes it so special. Then, we are going to work through the photographic process – from creating your own 9 mm wide film to scanning your pictures into a computer – and we'll discover how we can use 3D-printed components at every stage. There are full details of all the references in the repository for this article which can be found here: **hsmag.cc/MinoxResources**.

Figure 1 (opposite) shows a Minox B camera along with a film cassette and a pair of glasses, to show the size of the camera. If you were found carrying a Minox 60 years ago, it would be assumed that you were a spy, because you probably were. If you were seen wearing those glasses, they might well decide you were a spy too. Minox cameras were used extensively by both sides in the Cold War. The camera can produce extremely detailed pictures, and is easy to use and conceal. James Bond used a Minox camera in the 1969 film *On Her Majesty's Secret Service*, although he notoriously held it upside down – probably because this looked cooler. Even the chain fitted to the camera comes equipped for spying. The beads on it correspond to focus distances which you can set on the lens. This makes it easy to take super-sharp pictures of things (for example, secret documents) just by holding the bead on the subject and pulling the chain tight to set the distance.

The Minox B camera is entirely mechanical, although it contains an electric light meter powered

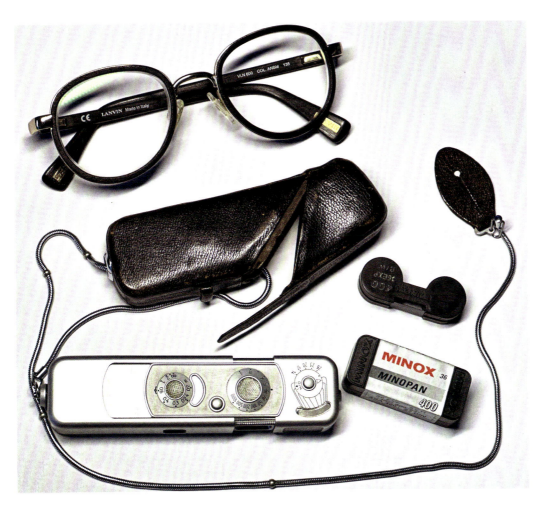

by a selenium light detector you can see on the right-hand side of the camera in **Figure 1**. There is a tiny needle in the light meter window above the detector. You adjust the right-hand shutter speed dial until the arrow matches up with the needle and your exposure is set. The left-hand dial sets the focus distance. You can set the focus so that everything further than 8ft away will be sharp. If the light is too bright, the camera has built-in filters you can slide in front of the lens to darken the image. If you want a Minox camera of your own, they can be picked up for around the price of a good-quality video game. →

Figure 1
The camera has been extended to its 'open' position to expose the lens. When the camera is closed, the film will be wound on to the next frame

There is a tiny needle in the light meter window above the detector

YOU'LL NEED

- Minox microcamera that uses Minox microcassettes
Search your favourite auction site for 'Minox B'

- 3D printer (preferably one that can take a 0.2mm nozzle)

- Some black 3D printing filament

- Some 35mm black and white film, and empty 35mm film cassettes

- A film changing bag (where you can load and unload the cassettes)

- A developer tank and a film spiral

- Access to a film scanner

TUTORIAL

Figure 2 ◈
The Minox camera has been used to take super-secret pictures of
Hull's The Deep aquarium and the Hull tidal barrier...

Figure 3 ◈
The biggest problem with small negatives is keeping
dust off them

AUCTION TIPS

The author has bought many items (including Minox cameras) from 'the world's
favourite online auction site', and considers himself familiar with the ways of buying
and selling things (mainly cameras). He therefore presents these handy tips for anyone
thinking of venturing online to get some gear.

1. Buy things for fun, not profit. It is hard work to buy cameras and make money from them.
   Much better to think of it as a hobby you are spending a bit of cash on. Many years ago,
   the author used to spend quite a lot of time and money on video games. These days, he is
   more likely to spend the price of a video game on an old camera and then have a bit of fun
   trying to make it work. If things end badly, he puts the camera back on sale with a humorous
   description of the problem, and usually gets some of his money back.
2. Don't get carried away when you are bidding. You will receive messages such as 'Don't let it
   get away' or 'It could still be yours' when you make a bid. Ignore them. Get a feel for the 'right'
   price and don't go beyond this. Unless you are bidding for the one and only Holy Grail (which
   is unlikely), there will always be another one coming along.
3. Beware of silly priced items. Listing an item for sale costs very little. Some sellers create
   listings with prices way above the market rate to try and drive up the market. You can use the
   'show only completed sales' filter to see how much items are really selling for.
4. 'Untested' usually means broken. If you buy an item described as untested, you are betting
   that you are smart enough to test (and then perhaps mend) what you are going to get. The
   author has lost this bet many times.
5. 'Tested' can also mean broken. Some sellers regard 'camera clicks when button pressed'
   as a passed test. Read the item description very carefully. Sometimes a 'tested' camera will
   have a broken exposure meter or film rewind. Some cameras are sold as 'film tested' (with
   samples), which is the best kind of tested.
6. Look for established sellers. The author has met some lovely folks buying and selling stuff.
   Look for people with good feedback who have been around a while. And don't be afraid to
   ask questions about what is being sold. A genuine seller will be happy to help.
7. Shop prices can be cheaper than auction prices. For a long time, the author thought that
   buying via auction was the cheapest way to get something. This is not always the case.
   Specialist dealers can be competitive and may offer returns and warranties. Check out their
   stores as well as auctions.

THE SMALL PICTURE

The Minox camera produces tiny images on
photographic film which is just 9 mm wide. **Figure 2**
shows this size in comparison to 110 cartridge film
(16 mm wide) and 35 mm film (35 mm wide). You
could fit many Minox images onto a single 35 mm
picture. This does limit the amount of detail that
you can get, but the pictures are very usable, and it
makes film much cheaper.

Figure 3 shows a print from one of the Minox
negatives shown in **Figure 2**. The detail on the
image is impressive and would be much improved
if film with smaller 'grain' had been used. A
photograph is made of tiny particles of silver which
are created in the parts of the film that have been
exposed to light. These particles are the 'grains'
that make the image. The faster the film (i.e. the
more sensitive to light it is), the larger the grain
particles, and the lower the quality. The grain in
Figure 3 is quite noticeable. Using a lower-speed
film would have produced a more detailed picture.

You can take colour pictures too (although
processing these at home is more difficult). If
you want to have your films processed (and
your cassettes refilled), search for 'Minox film
processing' online. However, the author much
prefers to do everything himself. Let's find out how
to do it.

A SLICE OF LIFE

The first problem that you encounter when you want to use your Minox camera is that 9 mm wide film is no longer made. Fortunately, this problem is easy to solve. **Figure 4** shows a film slicer which can be used to cut 35 mm film into 9 mm wide strips. Film is wound from the cassette on the left to one on the right. On the way, it is split by three blades embedded in the block in the middle. You place a new cassette in the left-hand slot, pull out a length of film, and attach it to the spool inside an empty cassette you place in the right-hand slot. Then you put on the light-tight cover (this step is important) and turn the handle to pull the film from one cassette to another. When you have finished, you have a cassette full of sliced film. You can create your own film slicer from 3D-printable designs (**hsmag.cc/FilmCutter**), or you can buy a ready-made one (**hsmag.cc/MinoxCutter**).

Figure 5 shows the result of the slicing operation. The receiving cassette holds two lengths of 9 mm wide film. Now we have the sliced film, the next challenge is to get the film into the Minox cassette. This must be performed in complete darkness. The easiest way to do this is to use a film changing

bag. This is a light-tight bag with elasticated cuffs, allowing you to put your hands inside it without letting light in. The bag has a zip at one end which you open to put in the 35 mm cassette containing the sliced film and the empty Minox cassette. Then you close the zip, put your hands inside the bag, pull out a length of sliced film from the 35 mm cassette and roll it up tightly. Next, you put the rolled-up film into the feed chamber of the Minox cassette and put the lid on the feed chamber.

Figure 6 shows a cassette with an unexposed film in the right-hand chamber. The film to be exposed is pulled out of the cassette onto the take-up spool in the left-hand chamber. The final task is to stick the film to the take-up spool, put the spool inside the cassette, and then put the lid on the cassette. These steps can be performed in the light. We can then tape the lids to hold them in place (the best kind of tape to use is 6 mm wide Tamiya Masking Tape, which is normally used when painting models). Now we have a cassette that we can load into the camera and use to take some pictures. →

Figure 4 ◈
You can also get a slicing block that will cut a 16 mm wide film strip to be used in other miniature cameras

Figure 5 ◈
The author has been unable to think of a use for the strips of sprocket holes that are also created by the slicing process. Although, he did once load some into a Minox cassette by mistake

Figure 6 ◪
The right-hand chamber in the cassette holds unexposed film. The film is wound onto the spool in the left-hand chamber after each shot

MINOX HISTORY

The Minox camera was designed in the 1930s by a German engineer with the wonderful name of Walter Zapp. It was one of the earliest 'system' cameras. Along with the camera, you could get a special tripod to stand it on, an adapter for binoculars, a film developing tank, and even a briefcase-sized enlarger you could use to create prints of your images. In the 1970s, the Minox B was replaced with the Minox C, which added automatic exposure and was even smaller (although the lens was fixed focus, which reduced the image quality slightly). However, by the end of the 20th century, digital cameras were becoming cheaper and able to compete on quality, leading to Minox camera production ending in 1996. The Minox name lives on, but these days, it is more likely to be attached to a pair of binoculars.

QUICK TIP

When printing delicate items like Minox cassettes, it is a good idea to reduce the print speed to about a quarter of your usual speed.

Figure 7 ◆
The Minox tank is on
the right-hand side

Figure 8 ◥
Make sure to get rid
of all the dust on the
film and the holder
before scanning, just
like the author didn't

QUICK TIP

Before you try to
load a cassette in
the dark bag, it is
best to practise
a few times in
the light.

INTO THE SPIRAL

To develop your own photographs, you need a
developing tank. This is a light-tight container which
holds the film and the liquids that develop the
image on the film and then fix the film so it can be
viewed in the light. The Paterson brand is popular,
and its tanks have been available for many years
Search for 'Paterson Universal' tank. The tank is
supplied with a 'spiral'. You push your film into the
spiral and then load it into the tank. To process
Minox films, you need a spiral which can take the
9 mm wide film. You can find a design for one at
hsmag.cc/PatersonFit. You put the exposed film
and the developing tank into your light-tight bag and
pull the film out of the cassette and push it into the
spiral. Then, you put the spiral on the holder, pop it
into the tank, and put on the tank lid. Now you can
take the tank out of the light-tight bag and perform
the rest of the development in normal light.

  Figure 7 shows a Paterson tank, and a spiral partly
loaded with Minox film on the holder on the left. If
you wanted to develop multiple films at the same
time, you could print several spirals and load them
all onto the spindle. Just make sure that you use
enough developer to cover all the spirals. If you can
afford it, you can get another wonderful piece of
design from Walter Zapp in the form of the Minox

daylight developing tank shown on the right of
Figure 7. This lets you develop Minox cassettes
without needing a darkroom. The Minox cartridge
is placed in the light-tight compartment on the
right-hand side of the Minox tank. You then rotate
the cylinder sticking out of the top of the tank to
pull the film out of the cartridge, and wind round the
outside of the cylinder in the tank for development.
Once the film has been developed, it needs to be
dried in a dust-free environment (the author uses the
bathroom) and the next step is to scan the pictures
into a computer.

SCAN THE HORIZON

Now that we have our images, we must scan them
into a computer. We need a flatbed scanner which
has a backlight for scanning negatives. You can pick
up such scanners quite cheaply, but make sure that
you get the software drivers for them as well. In Hull,
UK, where the author lives, there is a local 'library
of stuff' (**libraryofstuff.co.uk**) where you can rent
things like film scanners (and lots of other things)
for very low prices. You might find one in your area.
You might also like to investigate local photography
clubs where they might have a communal scanner, or
someone happy to help you further your hobby and
scan some negatives for you.

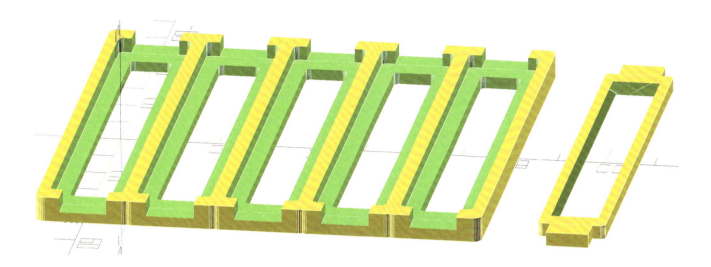

Figure 8 shows the film in a film holder, ready for scanning. Scanners are usually provided with negative holders, but you won't find any that fit Minox film. This is another problem we can solve with our trusty 3D printer.

Figure 9 shows a film scanner design produced by the author which is easy to print and use. Each of the holders is a slightly different height so that we can compare different film positions and then print a set of holders with the best possible focus. There are sets of holders available in the resources for this article.

I SPY WITH MY LITTLE MINOX

The author has had a lot of fun with his 'spy' camera. The complexity of producing film and creating pictures adds a lot to the photographic experience. You really feel like you own the shots that you end up with. The camera itself is a genuine marvel of design and construction. It doesn't look like something made over half a century ago. And, once you have had the experience of handling film in a light-tight bag and processing, you can work with other sizes and types of film. A tiny camera, like a Minox, can be the first step into a rewarding photographic hobby. And making it work makes very good use of other maker skills. ☐

> **The complexity of producing film and creating pictures** adds a lot to the photographic experience

WHERE DO WE GET MINOX CASSETTES?

Minox cassettes were last sold in the 1990s, nearly 30 years ago. Fortunately, there are still quite a few around, and you can find 'new old stock' Minox films. The cassettes can be reused, and you can even obtain metal versions, at a price. Search for 'Minox cassettes'. However, what we really want to do is 3D-print our own cassettes. There are some designs online, but these don't fit the author's camera. He has made a remix of a design with improved take-up spools. He is also working on a cassette which has a take-up spool on each end, to ease loading. Rather than having to wind the film into a small roll by hand, you will be able to just wind it directly into the cassette.

The cassettes can be printed on a normal 3D printer, but they require the use of a 0.2 mm printer nozzle (the standard printer nozzle size is 0.4 mm) because the walls of the cassette lids are very thin. The author is working on a cassette design which can be printed with a 0.4 mm nozzle. It won't be able to hold as many shots, but it will be easier to make. You will also need to add felt light traps to the slots in the cassettes – search for 'self-adhesive felt' and get some which are 0.5 mm thick. You can find all of the designs in the GitHub repository for this article here: **hsmag.cc/MinoxResources**.

QUICK TIP

The blades in the slicer are very sharp. Be careful to keep your fingers away from them when you are slicing film.

Figure 9
The design was produced using OpenSCAD. The code is in the repository for this project

Left
You should use black filament to make your cassettes, not the colour shown here

TUTORIAL

Remote control a camera with Raspberry Pi Connect

Dial in to your little computer, wherever it is

Ben Everard

Ben's house is slowly being taken over by 3D printers. He plans to solve this by printing an extension, once he gets enough printers.

Raspberry Pi Connect is a new service from Raspberry Pi that lets you connect to your Raspberry Pi from another computer on the internet. You sign in via the Raspberry Pi website and, from there, you can access the desktop session on your Raspberry Pi and control it as though you were sitting in front of it.

It's similar to how VNC works, with one big difference: Raspberry Pi Connect will help you route your connection through the internet. This means that you can connect to a Raspberry Pi on your home network when you're in a completely different place. Provided the Raspberry Pi and the computer you want to connect to it from are both on the internet, then a connection will get through.

We're going to use this to create a pet camera for keeping an eye on our pets.

To get started, you'll need a Raspberry Pi ID. You might already have one, but if not, you can create one at **id.raspberrypi.com**. Once you've got your username and password, you can set up your Raspberry Pi.

First, you'll need to open a terminal to install the relevant software:

```
sudo apt update
sudo apt upgrade
sudo apt install rpi-connect
systemctl --user start rpi-connect
```

The first two lines make sure that your system is up to date, the third installs the Raspberry Pi Connect client, and the final one starts it. At this point, you should see the Connect logo in the top right-hand corner of your screen. If you click on this, you can select sign-in.

This will open your web browser, and here you can enter the login details you created before. This will add your Raspberry Pi to your Connect account and now you can access it from anywhere.

On a different, internet-connected computer, head to **connect.raspberrypi.com**, and sign in. You should see a list of all connected devices – there should be the one you've just connected (and any others that you've previously connected). Click on Connect

GOING FURTHER

Perhaps the most useful extension to this would be to mount the camera on a pan-and-tilt arm so that you could move it around. Typically, these use two servos and brackets to let you move the camera in two axes.

There are a few kits available with this, or you can go the DIY route.

If you want to take things even further, you could add some way to interact with your pet. For example, you could mount a laser pointer on a servo so you can wiggle it, or perhaps move a string with a cat toy on it.

and a new window will pop open. If the Connect button isn't there, it means that the Raspberry Pi is connected to your device but not available to connect to. Make sure that you've signed in on that device and that it's powered on.

Once it's signed in to your Raspberry Pi – this might take a few seconds – you should see the desktop of that computer open in the web browser window. You can now control the mouse and keyboard remotely, just as if you were using a mouse and keyboard physically connected to the Raspberry Pi.

At this point, you can use the machine however you like. You might find this easy for helping friends and family with technical problems; you might find it an easy way to keep tabs on your home server. However, in this article, we're going to use it to create a pet camera. This is basically, a camera that we can set up at home and connect to keep an eye on our pets.

 Raspberry Pi OS comes with a few tools for working with the cameras

ON TEST

Our two test cats for this article are both kittens about nine months old, named Peach and Moon. Their welfare was a top priority while testing the software, and we can confirm that no cats were harmed in the making of this feature (though Peach did scratch the author, and Moon knocked a plant pot off the windowsill).

The first thing you need is a Raspberry Pi Camera connected to your board. It's safest to do this with the power off, so turn off your computer if it isn't already. You can now connect the camera cable to the camera port on a Raspberry Pi 4, or either of the camera/display ports on a Raspberry Pi 5.

Raspberry Pi OS comes with a few tools for working with the cameras, but they're all command line-based. There is nothing stopping you using these for the pet cam, and simply logging into the desktop and using the terminal. However, we'll be doing something a little more user-friendly.

We're going to use the Picamera2 WebUI Lite interface created by James Mitchell (aka monkeymademe on GitHub). This runs a web server that serves up a web page that lets you view and control the camera. →

Above Left
A Raspberry Pi ID can be used for various activities

Above
This pair are not as innocent as the image may suggest

This highlights the problem that Raspberry Pi Connect solves. After all, you may think, if the Raspberry Pi is serving up a web page, surely we can view this as we can view other pages on the web? The answer is usually (but not always) no, and for reasons that aren't particularly straightforward.

The first problem is addressing. You can send a message to any computer on the internet with an IP address. These addresses can either be version 4 (which is a series of four numbers separated by dots), or version 6 (which are alphanumeric strings separated by colons). For most of the history of the internet, version 4 has been the most popular. Given that IPv4 addresses are 32 bits long, there are just over 4 billion possible addresses. In a world with 9 billion people, where many people have lots of devices connected to the internet, this is a problem. One of the solutions to this is Network Address Translation, or NAT. This is where a local network (such as your home or office network), assigns a set of private IP addresses. These often start 192.126

or 10.10. There can be thousands, or quite possibly millions of devices with the IP address 192.168.0.1, and it's absolutely fine because they are only accessible on their local network. Computers outside of this can't send them messages.

Computers on these private networks can send messages to the wider world through the magic of NATs.

There are solutions to this, but typically they involve changing settings on your router, and this is a bit technical, and can potentially open you up to security problems.

Raspberry Pi Connect takes care of this headache, and gives you the desktop via the internet. We can use this desktop to access the web page running locally. Let's take a look at how to do this.

You need to run the following commands on the Raspberry Pi, but it makes no difference if you run them using a keyboard and mouse physically attached to the Raspberry Pi, or on another computer via Connect. Either way, open up a terminal and run

Above ◈
Raspberry Pi OS running in Firefox on Windows

Right ◈
You can tidy everything up with a 3D printed case

the following:

```
git clone https://github.com/monkeymademe/
picamera2-WebUI-Lite.git
cd picamera2-WebUI-Lite
python app.py
```

You can now open a web browser in the Connect desktop, and point it to **localhost:8080**.

You should see the output of your camera, and there are controls for managing it.

The problem with this is that the web server will stop if you restart the computer. You can set it to automatically restart by running the following command:

```
crontab -e
```

This opens the configuration file for Cron, which is a bit of software that handles regular operations. Select to use Nano (if you haven't used this before), then scroll to the bottom, and enter the following:

```
python /home/ben/picamera2-WebUI-Lite/app.py
```

…replacing 'ben' with your username. Press **CTRL+X** to save and exit. The camera server will now start automatically if you restart your computer.

SECURITY CAMERA

Our system works well for a simple, remotely viewable camera such as a pet camera. However, there are far more powerful options out there if you want more security camera-type features. For example, motionEye lets you perform particular actions when it detects movement in a frame.

Do be aware that if you are recording people, you will need to follow local laws on data protection.

FINAL STRETCH

That's the software set up. The last thing we need is to tidy up the hardware. We don't want to leave a bare Raspberry Pi sitting around the place, or it'll get damaged, and we need a way to mount the camera.

There's a huge range of options for both of these, and you may already have the bits you need. We opted to print a Raspberry Pi Camera Holder from the Raspberry Pi account on Printables (**hsmag.cc/camholder**), and a Raspberry Pi 5 Case by Hasan Yildiz (**hsmag.cc/5case**).

With these, we can set up our pet cam wherever we want, and keep an eye on this pair of mischievous felines. ▫

3D scanning with your phone

Trap real objects in your own virtual world

Above ◈
This two-colour print was
done on a Prusa XL

Ben Everard

Ben spends more
time making games
controllers than he does
playing games. It's a
strange hobby, but he
enjoys it.

W

e live in a 3D world, and are surrounded by thousands of 3D objects every day. We can use a bit of digital magic to turn these into 3D models. Basically, this involves taking a large number of photographs from every conceivable angle and running these through some clever software that creates a 3D model that we can then use just like any other 3D model. In this article, we will look at printing it out, but you can also use them as assets for games, or other 3D renders. This process is known as photogrammetry. There are some bits of software you can use to do the processing on a home computer. However, they typically require a GPU, and often an NVIDIA one. We don't happen to have one of these in the workshop, and we suspect that many of you don't either. A far easier option is to use an online service which will do the processing for you. We used the KIRI Engine, which is available on the web or as an Android or iOS app. There are lots of options available, and they generally do one or more of the same basic processes.

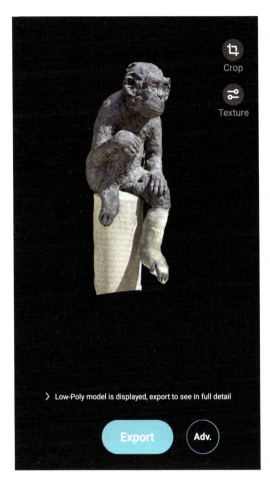

In KIRI, these are known as…

- **Basic Scan:** This is where you take a lot of photographs of an object, and the software attempts to build a 3D model from them. On KIRI, you can add 70 photos per project under the 'Basic' plan, which is free of charge, or 200 under the 'Pro' plan. Our experience was that 70 resulted in fairly poor scans, and you needed to go well over 100 to get good results. You can take the photos in the app either manually (i.e. pressing the shutter button each time) or automatically (where you move your camera around and it takes the photos as needed). Both worked well. Under the Pro plan, you can also upload photos from another camera – this may be particularly useful if using a drone, or a fancy camera.

- **Featureless Object Scan:** With a standard scan, features on the object are used to identify particular points. This works well for many objects, but if your object is a plain colour or has smooth sides, you may find you get better results with a featureless object scan.

RUNNING LOCALLY

Cloud services are a great way of getting up and running quickly with minimal hardware or software installation. However, sometimes you want the extra control you can only get from running your own software on your own hardware. There are a few options for this, ranging from the hobbyist to high-end industrial. Here are our picks to help you get started for free:

- RealityCapture. Currently free for businesses or individuals making less than $1m per year as well as all students and universities. It's a powerful piece of software that can render accurate 3D scans quickly.

- Meshroom is an open-source photogrammetry pipeline. It is perhaps not quite as powerful as RealityCapture, but still makes great scans.

  Both of these require an NVIDIA graphics card. If you don't have one of these, you can get some of the power of running locally by using a Google Colab sheet, such as the one for Meshroom here: **hsmag.cc/Meshroom**.

- **Gaussian Splat:** Most scans start with a series of photos, but Gaussian splats instead start with a video. To be perfectly honest, we don't understand the maths behind this. Most of the time, Gaussian splats output a rather obscure file format that's a point cloud, and it pretty hard to work with. The KIRI Engine does allow you to output a mesh; however, you have to check the box at the start. You can't change this once you've already created your splat. If you use the in-app camera, it will tell you if you're moving too fast. However, you can also upload videos taken elsewhere. →

Above
Like this lion, this author also has a broken wrist, and took this scan while waiting for a medical appointment

Left ◈
This monkey (like the lion) is found outside Southmead Hospital in Bristol

BOOK OF MAKING 2026   83

Right ◆
You can crop the
models in KIRI, but
we prefer to do it
in Meshmixer

Whichever option you use, you'll typically leave the object still and move the camera. If you want to move the object instead, you need to check 'Auto-mask object'.

If you're a pro user of KIRI (or, if your other service allows it), we'd recommend taking the photos or video outside of the app. While the app is useful, if you have an offline copy of files, you can edit them, remove some, add more, try another method, or anything else you want to do later. If you use the app, you only get one chance to get the photos right and you have to upload straight away.

PICK A METHOD

Gaussian splats are by far the quickest method (at least as far as taking the photos is concerned – processing the data can take a while). If you're on the go, it'll also use far less network bandwidth to upload. The results are really impressive: we found that while they may have missed a bit of detail, they had far fewer anomalies than the photo methods, and as such, often required little more than cropping before they could be printed.

The photo-based scans, on the other hand, were a more mixed affair. They were prone to spurious holes and appendages. The more photos you have (and the better quality the photos), the fewer issues you'll have. Our experience was that if you had enough high-quality photos, the results were better than Gaussian splats, but if you didn't, then they weren't.

TOUCHING UP

There are a few reasons you might want to scan something, and your motivations will inform what you do with the results. Many scans are for scientific or illustrative purposes. In these cases, you probably want to do as little as possible after the scan. If there are errors, you might want to add more photos, or start again entirely.

However, if you're 3D-scanning because you want an interesting scan, or a fun thing to 3D-print, you might be willing to take a heavy hand in editing the resulting model to fix errors, or make it more printable.

WHAT CAMERA?

You can get started with 3D scanning using a smartphone. However, a better camera will result in a better scan. Low light performance can be an important factor because many things you want to scan will have nooks and crannies that you have to try and photograph.

Image size is obviously important, but bear in mind that you'll be taking lots of photos, so beyond a certain point it can be tricky to manage the data and take a long time to process.

Some photogrammetry software can take RAW format photos, and you may get a better-quality scan with these if your camera supports it.

You can reach for 3D modelling software for this. Blender, for instance, has powerful tools for editing 3D meshes. At least, we believe it does – this author has never mastered this particular bit of software.

Meshmixer is a more straightforward tool for working with 3D models. It's quite limited, but the range of tools works reasonably well for tidying up 3D scans. However, there is a slight problem. Autodesk has bought the software, and stopped giving it updates. It still works, but at some point in the future, it'll probably stop working. For now, though, it's our preferred option.

Our method for touching up 3D scans in Meshmixer is…

- **Step 1:** crop out unwanted parts. The chances are that your scan took in more than just the object you wanted. Not only are these extra bits a distraction you don't need, but they can be a huge number of polygons that your 3D software needs to keep track of. Getting rid of them speeds everything up and generally makes it easier. To do this, go to Edit > Plane Cut. This tool lets you position a plane in the 3D space and discard all the polygons on one side of it. It might take quite a few plane cuts to get rid of all the bits you don't want. If you find that a plane cut isn't working, then you might want to consider the physical setup for the scan. However, there is also a sphere cut in KIRI that you can use if needed.

- **Step 2:** make solid. A 3D scan doesn't create a solid output, it creates a surface. Hopefully, this surface is reasonably complete, but there are probably at least a few holes in it. Before we can use it for 3D printing, we need to complete this mesh and turn it into a solid object. To do this, go to Edit > Make Solid. There are a few options here – we find that Meshmixer is a bit too conservative with the mesh size and so we generally increase it to allow more detail to be retained. Slide the Mesh Density slider up or down and click Update to change the preview. When you've found the right balance between detail and complexity, press Accept. You usually find that at this point the model turns grey. This is because the texture data is lost. In 3D terms, texture isn't texture at all. It makes no difference to the surface of the object – it's just 2D images that are mapped onto the object. This means that as far as 3D printing is concerned, it's irrelevant.

- **Step 3:** fix errors. The sculpting menu gives you some tools to fix any problems you have in the mesh. It takes a bit of practice to work out what tools can solve what problems, so we'd recommend allotting a bit of time to practise, and save a lot as you go along, as it's easy to get into a position where you've almost got it how you want it, then suddenly getting it all wrong.

Left
Make sure you export in a format you can import into your slicer – we chose OBJ

SLICING

Once you're happy with the results, it's time to bring it into your slicing software. We use PrusaSlicer for this, as it makes it easy to fix the last few bits. If your scan includes any regular shapes – such as cubes or cylinders – you can replace them with imported versions of the same shapes that have sharper lines.

Highlight your model and right-click, then select Add Part, and then the appropriate shape. You can use the usual move and scale tools to make the part the right size.

You'll probably find that the scan isn't a particularly easy shape for 3D printing. Most real objects have thin bits, overhangs, and all manner of things we're told to avoid in our 3D-printable designs. Fortunately, modern 3D printers and slicers are much better than they used to be, so we don't need to worry too much about this. We have found that our 3D scans print a bit better if we use the automatic support painting feature of PrusaSlicer. To use this, highlight your model, then click on the paintbrush icon (or press **L**), and press Automatic Support Painting. In the Supports section on the right-hand side, select For Support Enforcers Only. This should ensure that enough bits are supported, but there's not too much filament wasted, and it's easy enough to remove the supports from the finished print.

Once you're happy, slice as usual, and send the file to your printer. Photogrammetry is easy to get started with and you can create some really interesting parts for 3D printing. ◻

Make a retro telephone which answers your questions

Add voice input to a Raspberry Pi-powered telephone and create a large language model telephone exchange for the phone to talk to

Rob Miles

Rob Miles has been playing with hardware and software since almost before there was hardware and software. You can find out more about his so-called life at **robmiles.com**.

Build on a Raspberry Pi-powered telephone to add voice input. Then create a Raspberry Pi 5-powered large language model (LLM) telephone exchange for the phone to talk to.

Figure 1 shows the author's 'Red Telephone'. To the left of the phone is 'The Exchange', a device which uses an LLM to answer questions. The user can pick up the phone, dial a number, ask a question, and put the phone down. After a while, The Exchange will call back with an answer which may not be correct, or even rational, but is always interesting. You can find all the construction details for the phone and The Exchange in the GitHub repository for this project at **hsmag.cc/Pi_Phone**.

CARBONATED AUDIO

Figure 2 shows the microphone in the telephone handset. The microphone is packed with carbon particles which vibrate when hit by sound waves.

The vibration causes the electrical resistance of the microphone to change and was used to produce a signal to be sent over the phone lines. Unfortunately, this kind of microphone is incompatible with modern audio inputs which require a voltage from a microphone, not a changing resistance. The first step in the project was to create a circuit that allows the microphone to be used with the USB audio adapter connected to the Raspberry Pi in the phone.

The author is quite proud of the circuit in **Figure 3**. The carbon microphone is connected to the left-hand connection and the microphone input to the right. You can speak into the handset and record the resulting audio on the Raspberry Pi. The circuit contains a 'potential divider', which is a posh name for some resistors wired in series across a potential difference. The potential difference in **Figure 3** is the 5 volts from the power supply. The resistors are R1 and the carbon microphone. One of the magical features of electricity is that a potential 'spreads itself out' across

Figure 1
The Exchange contains an 8GB Raspberry Pi 5 with a 256GB SSD. The telephone contains a Raspberry Pi Zero

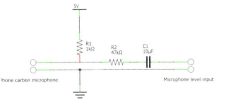

fritzing

Figure 2
The blue and white wires are connected to the microphone. The green and red wires are connected to the speaker

Figure 3
Increase the value of R2 to reduce the microphone output signal

YOU'LL NEED

- **Raspberry Pi 5**
  with 8GB of memory

- **A 3D printer**
  (preferably one that can take a 0.2 mm nozzle)

- **NVMe Base with an SSD card**
  (the author used one from Pimoroni)

If you are adding a microphone input to the Raspberry Pi-powered telephone:

- **2 × resistors**
  (1kΩ and 47kΩ)

- **10 µF capacitor**

- **A small piece of circuit board**

---

the resistance in a circuit. The 5 volts across the potential divider in **Figure 3** is spread across the circuit in a manner proportional to the resistance values. The total resistance is 1,500 Ω (1,000 Ω for the resistor + 500 Ω for the microphone). The 1kΩ resistor is two-thirds of this total, so two-thirds of the voltage is 'dropped' across this resistor. The point where the 1kΩ resistor and the handset microphone are connected should, therefore, be at a voltage of one-third of 5 volts (1.66 volts), the other two-thirds of the potential having been dropped across the 1kΩ resistor.

So, we have a potential divider which contains two resistors: the 1kΩ one and the carbon microphone. When I speak into the microphone, the carbon granules vibrate and the resistance of the microphone changes. This changes the voltage at the point where the microphone and the fixed resistor are connected, generating an electrical signal that represents the sound.

The signal goes through a resistor (R2) to reduce its level and then into a capacitor which only lets through the alternating current (the sound signal). This is connected to the microphone input of the USB audio adaptor plugged into the Raspberry Pi inside the phone, and hey presto, we have audio.

Figure 4, overleaf, shows the completed circuit inside the phone. The blue and white cables with the spade connectors are the input signal; the green and yellow wires at the bottom of the board are the output. The red and green wires at the top are the input voltage. The leftmost track on the circuit board →

Figure 4 ◈
The author has
spared you the sight
of the soldering
underneath the board

(the one with the green wires in it) is connected vertically underneath the board; all the other tracks are connected horizontally. It makes sense if you stare at it long enough. And the author has found that it works. The sound quality is not as good as you get from a modern dynamic or condenser microphone, but it is good enough for the voice recognition software we are using. And makes for an authentic telephone voice sound.

SPEECH-TO-TEXT

Now that we have an audio input for our phone, the next thing to do is add some software to convert recorded speech files into text. There are lots of speech-to-text converters available. Some send the sound information into the cloud, where a powerful computer performs the conversion. The author was keen to perform all processing inside the phone and found a library at **hsmag.cc/spchcat** which runs on a Raspberry Pi Zero, albeit a bit slowly. The **spchcat** program is used from the command line. You specify a file (or a sound source), and the program displays the speech that it finds.

```
spchcat --json message.wav > message.json
```

The statement above tells **spchcat** to read the sound recording in the file **message.wav** and create an output file called **message.json** describing any speech that was detected. We can run this from the console, but we really want to want to run it inside the JavaScript program that is controlling the phone. The first thing we do is put the command into a JavaScript string variable:

```
const decodeCommand = `spchcat --json message.wav
> message.json`;
```

The string **decodeCommand** holds the command we want our running program to execute. We can use the **exec** function to run this. It executes a command and then calls a JavaScript function when the command is finished. The **exec** function is a member of the **child_process** module.

```
const { exec } = require('child_process');
```

The statement above imports **exec** into the phone application. The **exec** function accepts a string containing the command and a reference to a function which accepts three parameters – an error message and the output and error streams. The function is called when the command completes (i.e. after **spchcat** has finished running).

```
exec(decodeCommand, (error, stdout, stderr) => {
  if (error) {
    console.error(`Error: ${error.message}`);
    this.decoding=false;
  } else {
    let jsonMessage = fs.readFileSync(`message.
json`, 'utf8');
    let text = this.
decodeSpeechfromJSON(jsonMessage);
    this.decoding=false;
    this.owner.speechDecodedSuccessfully(text);
  }
});
```

The code above shows how **exec** is used to decode the speech. If the **spchcat** command works (i.e. there is no error), the code reads the output file that was produced by **spchcat** and calls the function **decodeSpeechfromJSON** to extract the words from this file. It then tells the owner that the speech was decoded successfully.

LISTEN WITH CONFIDENCE

The **spchcat** program can produce a JavaScript Object Notation (JSON) file which describes what it thinks it has heard. A JSON file contains a text description of an object. In this case, the object contains data fields including a confidence value and a list of the words that were recognised.

```
{
"metadata":{"confidence":-6.96942},
"words":[
    {"word":"how","time":1.24,"duration":0.42},
    {"word":"are","time":1.78,"duration":0.2},
    {"word":"you","time":2.08,"duration":0.04}],
}
```

The phone program presently doesn't use the confidence value, although this could be used to reject speech that might be indistinct. The **words** array contains the words that were detected, along with each word's duration and position in the sample. The function **decodeSpeechfromJSON** converts this array into a string of text.

```
decodeSpeechfromJSON(jsonMessage){
  const message = JSON.parse(jsonMessage);
  const text = message.words.map(w => w.word).
join(' ');
  console.log(`Text decoded successfully:
${text}`);
  return text;
}
```

The function parses the JSON string and creates an object that contains the data described by the JSON. It then uses the `map` function provided by the `words` array to extract the word strings from the array and put them into a single string separated by spaces. It displays the decoded text to the console and then returns the text it found. We now have a telephone that can listen to audio and extract spoken text from it. Now we must discover how we can use this to add features to the telephone.

AFFAIRS OF STATE

Figure 5 shows the commands supported by the phone. The original phone software only had two commands. The new phone has much more complex behaviours, and so it was decided to use a state machine to manage it. State machines are worth knowing about. At any given instant, the phone is in a particular state, waiting for events that will cause it to perform actions and then move into a different state.

Figure 6 shows the states used to manage what happens when the user performs a command. The initial state, `REST`, is at the top of the diagram. This state is connected to other states by events which fire at particular times. If the handset is picked up, the phone moves into the `DIAL_TONE` state. You can work your way through the way that the phone is used by examining the states and the events that move them from one state to another. All the states are held in a `stateActions` object.

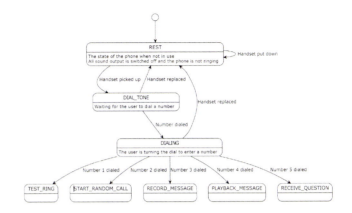

```
Exchange Instructions
Lift the handset and dial the following numbers to use our services:
    1    The telephone will ring until you replace the receiver
    2    Replace the handset. The telephone will ring later. Lift
         the handset for a message.
    3    The Exchange will answer. Speak your message and it will
         be stored.
    4    The Exchange will answer and play back your message.
    5    The Exchange will answer. Ask your question and replace
         the handset. The Exchange will call you back with an
         answer to your question.
Always replace the handset when your call is complete.
```

```
this.stateActions = {
    REST: {
        'Handset picked up': () => {
            this.ringer.ding();
            this.soundOutput.playFile('./sounds/
dialTone.wav');
            return 'DIAL_TONE';
        },
        'Handset replaced': () => {
            this.soundOutput.stopPlayback();
            this.ringer.ding();
```

Figure 6 ◆
You can use the website at plantuml.com to create diagrams from text descriptions of states and events

```
            return 'REST';
        }
    },
    // Other states go here
}
```

The code above shows the `stateActions` object which represents the `REST` state of the phone. When an event fires, the phone will find the current state and then call the function in that state which deals with the event. The function returns the new state for the phone. If the handset is picked up, the event handler makes the ringer go 'ding' – just like an old phone does – and then plays the dial tone and moves the phone into the `DIAL_TONE` state. The `REST` state doesn't contain handlers for all possible events; for example, turning the phone dial has no effect if the phone is in the `REST` state. Using states makes it easy to change the way the phone works and add new features.

You can use the telephone on its own. It can record and playback messages and respond to requests from the local network. However, it gets more interesting if connected to The Exchange. If you dial 5 on the phone it will record your spoken question, convert it into text and then send it to The Exchange for processing.

AN LLM EXCHANGE

Figure 7, overleaf, shows the insides of The Exchange. The case design has been modified to include the printed text on the side. You can find the STL file for the case in the repository for this project. The Exchange is a Raspberry Pi 5 with 8GB of memory and a solid-state drive (SSD). It runs an LLM that can create answers to questions. So, what does an LLM do? →

QUICK TIP

Small changes to the prompt string can make a big difference to the way the chatbot behaves.

Figure 5 ◆
Typing the instruction card was great fun

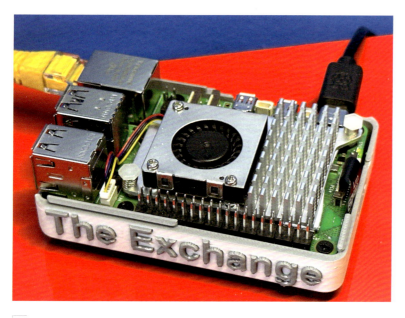

Figure 7 ⬆
When you ask it a
question, the fan
comes on

Figure 8 ⬇
'Bob' is one of the
chatbots supplied
with the LLM

QUICK TIP

The author has
not seen any bad
or rude behaviour
from The Exchange,
but there is no
way he would let
his seven-year-old
granddaughter
loose on it on
her own.

MODEL BEHAVIOUR

We'll leave it to philosophers to decide what
constitutes intelligence and whether or not large
language models have it. Essentially they are
probabalistic in that it looks at the input, and is
trained on large amounts of existing text and tries
to decide what is the most likely next bit of text. Is
this thinking? Does it matter?

An LLM model is a file containing things which
can be matched against inputs to generate outputs
and a program that does the matching. The model
files used by systems running in the cloud are
enormous, but we can obtain smaller ones that can
run on an 8GB Raspberry Pi 5 with an SSD. The
model file needs to be processed so that it can
be used on a Raspberry Pi – there is a link to the
procedure for doing this in the GitHub repository for
this article.

This model in The Exchange is around 4.5GB
in size. The Exchange server users the 'llama.
cpp' software to interact with the model. You can
find this at **hsmag.cc/llama_ccp**. You can use it
to create a 'chatbot' (a program that takes your
questions, sends them to the LLM, and then
displays the response).

TALKING TO CHATBOTS

Figure 8 shows a console conversation with the 'Bob'
chatbot running on the Raspberry Pi 5. He seems to be
pretty good on general subjects, although he got the
author of the *C# Programming Yellow Book* wrong.

PROMPT, PLEASE

When you use an LLM to create a chatbot, you need to
provide a 'prompt string' which is sent to the LLM with
every query to 'set the scene' for the query. It looks as
if you are providing the model with instructions that will
tell it how to behave, but what really happens is that the
chatbot finds words that match with the prompt as well
as the question you asked. This stops the LLM going off
on flights of fancy, something it is prone to doing.

```
let promptPrefix = `Transcript of a dialog, where
the User interacts with an Assistant named
Exchange. Exchange is helpful, kind, honest, good
at writing, and never fails to answer the User's
requests immediately and with precision.

User: Hello, Exchange.
Exchange: Hello. How may I help you today?
User: Please tell me the largest city in Europe.
Exchange: Sure. The largest city in Europe is
Moscow, the capital of Russia.
User:`;
```

Above, you can see the prompt string used for
queries to The Exchange. It tells The Exchange how to
behave and gives an example of a conversation. The
question to be asked is added onto the end of the string
which is to be sent to the LLM.

```
question = `${promptPrefix}${question}.`;

let req = {
    method: 'POST',
    body: JSON.stringify({
        prompt : question,
        n_predict: 256,
        temperature: 0.7,
        stop:["User:"]
    })
};
```

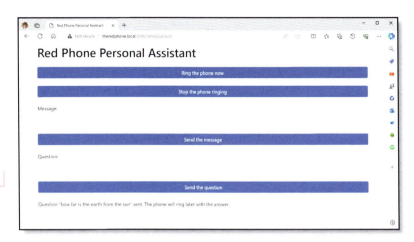

 We'll leave it to philosophers to decide what constitutes intelligence

Figure 9
The answer that came back was correct, which is nice

The code above creates a web request which is sent to a Chatbot server which is part of the llama.cpp distribution. The prompt prefix is added to the beginning of the question and sent to the server along with other setting values. The `n_predict` setting sets the number of predictions that will be used to build the answer, the `temperature` value sets how 'random' you want the answer to be, and the `stop` string prevents the chatbot from inventing its own questions from the user and answering them. As soon as the chatbot generates the string `"User:"`, it stops generating any more output.

EXCHANGE AND SMARTS

Figure 9 shows the Personal Assistant web page hosted in the phone. This can be used to send messages to the phone from a web page hosted in the phone. You type in the message, press 'Send the message', and the phone will ring immediately and playback the message when the receiver is picked up. The updated version of the phone allows you to type in a question and send it to the phone. The phone will then send the question to the LLM and then ring and speak the answer when it is received.

FUN WITH AN LLM

The service provided by The Exchange is not particularly good. It gets things wrong, doesn't always understand what you asked, and tends to go on about things which have nothing to do with what was asked. But the thing to remember is that this is all happening on small, low-powered devices with no assistance from external computers. You could use The Exchange anywhere, with no need for an external network connection. Furthermore, watching The Exchange get things wrong is actually very interesting, as is playing with the various settings and seeing what happens. The author hopes that you enjoy playing with this and that it improves your understanding of what the technology can and can't do. ☐

TELEPHONE HELPERS

Figure 10 shows all the 'helpers' in the phone application. You can use these in your JavaScript applications. There are helpers for hardware input-output, the LLM, and audio and speech input and output. The entire phone is controlled by an instance of a phone object. When the phone starts running, it creates instances of all the helpers it needs:

```
const SpeechInput = require('./helpers/speechInput');
```

The statement above brings the `speechInput` module into the program. A module contains a JavaScript object that you use to interact with the helper services.

```
this.speechInput = new SpeechInput(this);
```

The statement above creates a `speechInput` helper object for the phone. The helper is passed a reference to the phone object (that's what the `this` keyword means) so that the helper can send messages to the phone by calling methods on this object. When the phone wants to convert an audio file into text, it calls a method in the helper object and passes it the location of the file to be converted:

```
this.speechInput.startSpeechDecode(`./recordings/question.wav`);
```

The statement above shows how this is done. The code in `startSpeechDecode` in the helper object sends the `spchcat` command to start the speech decoding. When the helper object finishes decoding the speech, it calls a method in the phone to deliver the text that was decoded:

```
this.owner.speechDecodedSuccessfully(text);
```

The `owner` property of the speech helper is a reference to the phone it is working for. The phone can then do something with the text (perhaps send it to the LLM). This structure makes it very easy to add new behaviours (just add new helper objects) and also makes it easy to change how a helper works. If you wanted to use a different text-to-speech system, you would just have to change the contents of the `speechInput` helper, and the rest of the phone code would remain the same.

∨ helpers
- JS InGPIO.js
- JS LLM.js
- JS OutGPIO.js
- JS phoneController.js
- JS soundInput.js
- JS soundOutput.js
- JS speechInput.js
- JS speechOutput.js

Figure 10 ◆
You can use these in your projects just by copying these files into a 'helpers' folder in your applications

Programming an electronic brain

How the Robot Operating System helps you build complex robots

Figure 1 ◈
This robot is called the MasterPi and is made by Hiwonder: hiwonder.com

Figure 2 ◈
The USB connection on the right is for the camera on the robot arm

Rob Miles

Rob has been playing with hardware and software since almost before there was hardware and software. You can find out more about his so-called life at robmiles.com.

G etting into robot development has never been easier or cheaper. You can pick up a robot kit for not much more than the price of a video game. If you spend a little more money, you can get one with a robot arm on top, like the one shown in **Figure 1** above. Many of the robots use mecanum wheels, which allow the robot to move in any direction.

Figure 2 shows what is controlling the robot. A Raspberry Pi sits underneath a 'HAT' which manages the power supply (two 18650 lithium batteries) and the signals to control the motors and the robot arm. The robot runs a set of Python programs which control the hardware and allow it to perform preprogrammed tasks.

Figure 3 shows the remote-control application you can use to tell the robot what to do. The robot hosts a Wi-Fi access point to which you connect the mobile application. You can then select from pre-built behaviours. This works well, but what if you want to do more? The author was very keen to use his robot as a platform for learning the Robot Operating System (ROS). So that is what he has decided to do. The robot is presently driven by Python programs; the idea is to turn these into ROS nodes. But first, we must learn a bit about ROS itself.

COMING UP ROS-ES

An operating system is something you add to hardware to make it useful. Examples are Windows, MacOS, or Linux. The operating system takes the raw ability of the hardware (running programs, reading keyboards, saving data, displaying images on screens, etc.) and provides a user interface. The operating system lets you select the program you want to run. When you start the program, the operating system fetches the selected program from mass storage and then performs the instructions in the program. Some of the instructions will ask the operating system to do things; for example, a word processor will ask for a document file to be loaded into memory.

ROS takes the abilities of a computer system and makes it useful to a robot creator. ROS lets you break a system down into cooperating components. Components are created inside packages, which makes it easy to manage complex solutions. There are many pre-built components that you can incorporate into your solutions. ROS also provides tools you can use to express the physical design of your robot (or other mechanical system controlled by ROS elements) and simulate behaviour in a virtual (i.e. computer-generated) environment. ROS is a rich and complex system which can take a while to master. It provides effective solutions to robotic problems that we aren't even aware we have. Learning it will hurt your head a bit, but it is worth the effort.

INSTALLATION

ROS sits on top of the computer operating system and is closely coupled to it. This means that the versions of ROS you can use are determined by the operating system on your computer. You can run ROS directly on a Windows PC, but the installation is not for the faint-hearted as it involves compiling the ROS program source. It is much easier to use a Linux-based machine for which you can obtain compiled binary versions of ROS. On a Windows PC, you can use the Windows Subsystem for Linux

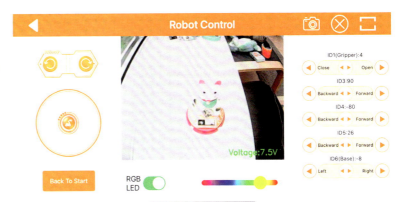

(WSL) to make a Linux environment into which you can install ROS. There is a step-by-step guide to the installation process in the GitHub repository for this article (**hsmag.cc/RosIntro**). The process uses the amazing Docker tool, which makes it possible to host any version of Linux on your machine, whatever its architecture or operating system.

ROS is a large application that needs plenty of memory to run. It will not fit on smaller devices as it needs at least 4GB of RAM. If you are using a Raspberry Pi, you can add a swap file to your system which uses file space to extend main memory. However, if you do this, you may find that your SD card (the place where your files are stored) wears out as the operating system will continually write to the swap file as programs are started and stopped. If you are feeling brave (and have a few spare SD cards), there are instructions on how to do this in the installation guide.

INSIDE THE WORLD OF ROS

ROS breaks a robot application down into nodes. A node is implemented as a running program. The nodes talk to each other in a well-defined way, and they cooperate to keep the robot running. The nodes can all run as individual tasks on a single computer, or they can run on lots of different processors connected to the local network.

Figure 4, overleaf, shows a ROS installation which contains a controller, a robot, and a camera. Perhaps we want to make a robot litter picker which will wander around searching for litter. The robot is constructed as three devices: a controller, a robot rover, and a camera. Each device is running one or more nodes that form part of the ROS solution.

Nodes provide services to other nodes and accept commands from them. In addition, any node can publish data items that any other node can subscribe to. There are lots of advantages to organising a robot application this way. It is easy to move nodes between devices. The connections between the →

Figure 3 ◆
This application is running on an iPhone. A version of the code is also available for Android

YOU'LL NEED

◆ **A reasonably powerful desktop computer, laptop, or Raspberry Pi (preferably a 4 or 5 with 4GB of RAM)** to run the development environment and ROS

◆ **Your own little robot or the parts to make one**

QUICK TIP

You can use your knowledge of ROS to power a pre-built robot or to take control of a robot that you have created yourself out of components you have chosen. There are tips on hardware choices in the GitHub repository for this article: **hsmag.cc/ RosIntro**

nodes are well-defined. We could move all the nodes onto a single powerful computer, or replace the vision system with one that uses a different camera.

We don't have to write the code for all the nodes. ROS also includes a library mechanism that makes it easy to import node code. To properly understand all of ROS, you must understand all the problems that it solves, and there are many of those. Let's start with something simple. How do ROS nodes help us get our 'litter picker' robot going?

CLEANING UP

To activate the litter-picking robot, the user could press the 'Search' button on the controller device. The buttons on the front panel of the controller are managed by the **buttons** node, which publishes the button states on a topic called 'buttons'.

The **manager** node has subscribed to this topic, so it receives a message informing it that the Search button has been pressed. The **manager** then sends a service request to the **vision** node asking, "Can you see anything?"

The vision system has subscribed to frames of image data published by the **camera** node and is looking for litter in each frame it receives. If the **manager** gets a response indicating that some litter has been spotted and giving a direction to that litter, the **manager** sends commands to the **motor** node in the robot to head that way. The **manager** also publishes the message 'moving' to a robot status topic.

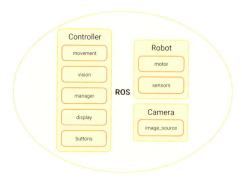

Figure 4 ◆
There are eight nodes in this system

The **display** node has subscribed to this topic, so the status 'moving' is displayed. As the robot moves, the **sensors** node on the robot would be publishing information about things the robot is detecting around it. ROS provides the environment in which all this would be made to work. We design the structure of the published data, write the code for each node, and then decide how the nodes

will interact, but ROS sits underneath and makes everything work.

PUSH THE BUTTON

We can discover how ROS does this by considering the very start of this process: the front panel of the robot. At the very least, the robot will have a button to make it start and a button to make it stop. It should also have a way of indicating what it is doing, perhaps a coloured light or a text display. The diagram in **Figure 4** shows that the Controller is running a **buttons** node which receives user commands and a **display** node that displays the robot status.

Let's look at the code in this node, which is implemented by a Python class called **ButtonPublisher**. This class extends the **Node** class which is part of ROS. An instance of **ButtonPublisher** class is created when the robot system starts up. Below, you can see the constructor method which runs when a **ButtonPublisher** is created.

```
class ButtonPublisher(Node):

    def __init__(self):
        super().__init__('button_publisher')
        self.buttonReader = ButtonReader()
        self.start_publisher()
        self.start_timer()
```

The constructor first calls the **__init__** method for the **Node** parent class, supplying the call with the name of the publisher. This tells ROS that there is a new node called **button_publisher** in town. The next statement in **__init__** creates a **ButtonReader** object which will be used to read the buttons on the physical console.

The **ButtonReader** reads the buttons from the controller hardware, perhaps by using GPIO pins (although it could also connect to a computer keyboard or touchscreen). Next, the constructor calls **start_publisher** to create a publisher to publish button events to anyone who is subscribed to them. Then it calls the **start_timer** method to start the button scanning timer. There is not much code in the **start_publisher** method:

```
def start_publisher(self):
    self.publisher_ = self.create_
publisher(String, 'buttons', 10)
```

The publisher to be used by **ButtonPublisher** is stored in a class member called **publisher**. The **create_publisher** method is inherited by

`ButtonPublisher` from its `Node` parent class. The `publisher` method accepts three parameters. The first indicates that the publisher will publish a String, the second gives the topic for the published message (in this case, 'buttons'), and the third parameter (the value 10) means 'keep the last ten published items and use the "RELIABLE" level of quality of service' – which means that published items are guaranteed to arrive at the receiver. The second method called by the `ButtonPublisher` constructor is `start_timer`:

```
def start_timer(self):
    timer_period = 0.1  # seconds
    self.timer = self.create_timer(timer_period,
self.timer_callback)
```

This method creates a timer which fires ten times a second. Each time the timer fires, the `timer_callback` method is called.

```
def timer_callback(self):

    button = self.buttonReader.scan_buttons()
    if button!="":
        msg = String()
        msg.data = button
        self.publisher_.publish(msg)
        self.get_logger().info('Publishing: "%s"'
% msg.data)
```

The `timer_callback` method calls the method `scan_buttons` on the button reader. This scans the buttons and returns the name of the button which is pressed, or an empty string if no buttons are pressed. If a non-empty string is returned, the callback publishes the name of the button that is pressed on the 'buttons' topic.

The final part of the button node program is the mechanism which starts the node itself. This is performed by the `main` method in the node code which is called when the node is loaded.

```
def main(args=None):
    rclpy.init(args=args)

    button_publisher = ButtonPublisher()

    rclpy.spin(button_publisher)

    # Destroy the node explicitly
    button_publisher.destroy_node()
    rclpy.shutdown()
```

The `main` method creates a `ButtonPublisher` instance and then calls the `button_publisher` method

on this. The `rclpy.spin` function keeps the node alive; responding to events and managing callbacks. It ends if the node is terminated by ROS, at which point the publisher node is destroyed and the node shuts down.

All the nodes in the robot are started in this way. The robot controller code can contain a launch method which starts all the nodes when the robot begins running.

SUBSCRIPTION MODEL

We know how a node running in a robot can post a message on a topic. Next, we need to consider how another node can receive it.

```
class DisplaySubscriber(Node):

    def __init__(self):
        super().__init__('display_subscriber')
        self.subscription = self.create_
subscription(
            String,
            'buttons',
            self.listener_callback,
            10)
        self.subscription  # prevent unused
variable warning

    def listener_callback(self, msg):
        self.get_logger().info('Button: "%s" was
pressed' % msg.data)
```

The code above defines a `DisplaySubscriber` class which subscribes to the `buttons` topic and calls the `listener_callback` function each time a message is received. This listener callback simply logs the button name (although it could put it on a text display). Code in the `manager` node could also subscribe to the `buttons` topic so that the manager is informed when a button is pressed.

PACKAGE HOLIDAY

Code for ROS applications is organised into packages. A package brings together a set of related behaviours. We could create a package called `front_panel` which contains the code for the `buttons` and `display` nodes.

Figure 5, overleaf, shows the package files for the `front_panel` package. The two highlighted files contain the code for the `buttons` and `display` Python nodes that we have seen above. The three files at the bottom of the package are what ties the package together. The **package.xml** file contains a description of the package and identifies any dependencies that the package has. →

QUICK TIP

The most recent version of ROS is called ROS2. This is the one you should be using.

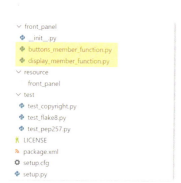

Figure 5 ◈
ROS can generate a template package for you to fill in

One package can use elements from another, and this is where you would identify the source packages. The **setup.cfg** file specifies where the files are to be placed, and the **setup.py** is a chunk of Python which is run to set up the project. This file is worth looking at:

```
from setuptools import find_packages, setup

package_name = 'front_panel'

setup(
    name=package_name,
    version='0.0.0',
    packages=find_packages(exclude=['test']),
    data_files=[
        ('share/ament_index/resource_index/
packages',
            ['resource/' + package_name]),
        ('share/' + package_name, ['package.
xml']),
    ],
    install_requires=['setuptools'],
    zip_safe=True,
    maintainer='rob',
    maintainer_email='rob@hullpixelbot.com',
    description='Provides a front panel containing
buttons and a text display',
    license='Apache-2.0',
    tests_require=['pytest'],
    entry_points={
        'console_scripts': [
            'buttons = front_panel.buttons_member_
function:main',
            'display = front_panel.display_member_
function:main'
        ],
    },
)
```

Above is the **setup.py** file for the `front_panel` package. The most important part of this file is the entry point list at the bottom. It specifies the Python files that are to be run when the nodes are activated. If we added a third node to the panel (perhaps a buzzer controller), we would create the Python code for the node and then add the entry point here. The above package serves as a template for any future packages you want to create. Just add the Python code into the same folder as the existing nodes and configure **setup.py** with the new entry points. Your code can then be run as a node in the robot system.

BUILDING THE PACKAGE

Now that we have our package, the next thing we need to do is build it. This is the process of collecting all the package components and getting them into a state where ROS can run them. Python programs are not compiled, but ROS nodes can also be created from C++ code, which does need to be compiled before it can run. A package that contains C++ code has a slightly different format and contains a **CMakelists.txt** file which describes how to build the code. It makes creating a package a bit trickier, but it does mean that you can use both languages in your solution. Furthermore, you can import packages containing nodes programmed in C++ and they will work alongside your Python nodes. The ROS system provides a command called **colcon** (collect components) that performs the build process:

```
colcon build --packages-select front_panel
```

The command above would be performed to build the `front_panel` and make the files ready to run. So, at last, we can run our robot nodes. There's just one more thing we need to know about (sorry), and that is all to do with sourcing our commands. We talk to the operating system from within a 'shell' environment which accepts our commands and then performs them. On Linux, this is usually the 'bash shell'. Some of the commands are 'built-in' to the shell; for other commands, the shell will go off and look for a program to run. We can tell the shell where to look for our robot node code:

```
source install/setup.bash
```

The **source** command means 'find this file and execute it as if it has been typed in'. The command file is called **setup.bash**, and it is created for us by **colcon** when the package is built. Once we have performed this command, we can use the `ros2 run` command to start our two nodes running:

Figure 6 shows two bash shells running the **buttons** and **display** nodes. The two nodes were

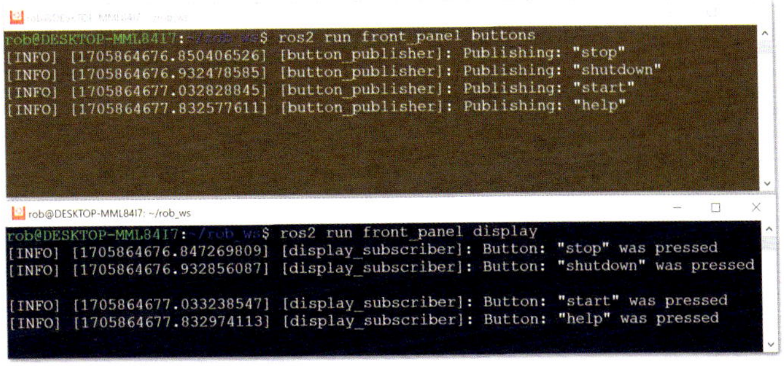

rob@DESKTOP-MML8417:~/rob_ws$ ros2 run front_panel buttons
[INFO] [1705864676.850406526] [button_publisher]: Publishing: "stop"
[INFO] [1705864676.932478585] [button_publisher]: Publishing: "shutdown"
[INFO] [1705864677.032828845] [button_publisher]: Publishing: "start"
[INFO] [1705864677.832577611] [button_publisher]: Publishing: "help"
rob@DESKTOP-MML8417: ~/rob_ws
rob@DESKTOP-MML8417:~/rob_ws$ ros2 run front_panel display
[INFO] [1705864676.847269809] [display_subscriber]: Button: "stop" was pressed
[INFO] [1705864676.932856087] [display_subscriber]: Button: "shutdown" was pressed
[INFO] [1705864677.033238547] [display_subscriber]: Button: "start" was pressed
[INFO] [1705864677.832974113] [display_subscriber]: Button: "help" was pressed

Figure 6 These packages are running on a PC under the Windows Subsystem for Linux, but they could just as easily be running on a Raspberry Pi

started using the commands at the top of each shell. The **`buttons`** node has published four button messages and the **`display`** node has displayed them.

ROS COMPLICATIONS

At this point, you might be thinking one of two things: 'Blimey, this is complicated' or alternatively, 'Blimey, this is powerful'. The complexity is there for a reason. We want to be able to break our robot controller down into smaller reusable components which can be easily swapped without affecting the rest of the application.

One robot might use a physical button to start it, but the next might have a touchscreen. Because this input is managed by a node, we just need to swap out the package node for another. Later, we might decide to combine the camera and the controller into a single device. In that situation, we just have to change the **`image_source`** node to run on the controller, and everything which uses images will just work.

The package folder contains everything needed to build and run a particular set of robot nodes. There are many pre-built packages for robot interfaces and behaviours that you can slot into your robot and interact with via their services and topics.

NODE MANAGEMENT

One of the many wonderful things about ROS is the way that things 'just work'. If you want two nodes to communicate, you just run them and they can magically see each other.

ROS works over your local area network (LAN). Network communications are based on the DDS (Data Distribution Service) standard, which provides the discovery protocol by which nodes can identify themselves and find other nodes. ROS provides commands we can use to discover nodes, topics, and services.

Figure 7 shows the output from three ROS commands which can be used to view an active ROS installation. The **`node list`** command lists all the active nodes – in this case, the **`button`** and the **`display`**. The **`topic list`** command shows all the active topics. The **`buttons`** topic is active, along with two which are provided by ROS. The final command shows all the services that are provided. Note that the button and the display components expose a set of services which can be used to work with them. This makes it possible for a ROS application to automatically discover what nodes can do and how

/button_publisher
/buttons

Figure 8 You can add plug-ins and lay out the display as you like

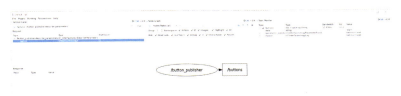

to use them, which makes possible self-configuring systems. We can see this in action if we start up the rqt tool, which is supplied as part of ROS. This contains a range of plug-ins you can use to create interfaces with your ROS application.

The rqt window in **Figure 8** shows a view of the topics available on the left, a diagram showing the active robot topics in the middle, and a log of button topic messages on the right. We can see that the last message that was sent was the 'start' message. These views are all updated in real time.

Now we know how ROS applications are structured and how the components interact. In the next article, we'll build on this. □

rob@DESKTOP-MML8417:~/rob_ws$ ros2 node list
/button_publisher
/display_subscriber
rob@DESKTOP-MML8417:~/rob_ws$ ros2 topic list
/buttons
/parameter_events
/rosout
rob@DESKTOP-MML8417:~/rob_ws$ ros2 service list
/button_publisher/describe_parameters
/button_publisher/get_parameter_types
/button_publisher/get_parameters
/button_publisher/get_type_description
/button_publisher/list_parameters
/button_publisher/set_parameters
/button_publisher/set_parameters_atomically
/display_subscriber/describe_parameters
/display_subscriber/get_parameter_types
/display_subscriber/get_parameters
/display_subscriber/get_type_description
/display_subscriber/list_parameters
/display_subscriber/set_parameters
/display_subscriber/set_parameters_atomically
rob@DESKTOP-MML8417:~/rob_ws$

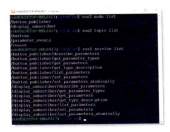

Figure 7 If other ROS nodes were active, they would appear in this output, too

Get a ROS-powered robot moving

Create a drive system for a ROS robot with motor control and movement feedback

Above ◈
The motion sensor clips underneath the robot and is connected via a Raspberry Pi Pico

Below ◈
You can see the controller HAT at the top right-hand corner, second one down

Rob Miles

Rob has been playing with hardware and software since almost before there was hardware and software. You can find out more about his so-called life at **robmiles.com**.

I n the last article, we discovered how ROS 2 applications are structured and how components can communicate using publish and subscribe. In this article, we are going to build on this to make the robot move, and then we are going to discover how we can add a sensor to determine the distance and direction of movement.

ROBOT CONTROL

The robot kit is from Hiwonder (**hiwonder.com**). It uses a HAT which fits on the Raspberry Pi driving the robot. The HAT drives the motors and servos and is controlled by Python code supplied by Hiwonder. It works well, but the author was keen to use ROS 2 with the robot, so the first step was to convert the motor driving code into a ROS 2 node with the original name of motors.

The motors node will subscribe to a topic which will be used by any process wanting to tell the robot to move. However, before we can create the motors node, we must look at how the motors are presently controlled by the Hiwonder code.

CONTROL POWER WITH PYTHON

Each of the wheels on the robot is turned by a DC motor connected to the controller HAT. The controller HAT creates pulse-width modulation (PWM) signals to control the power going into the motor. PWM works by rapidly turning the driving voltage on and off to vary the amount of power being delivered. You can get the effect of a dimmed light at home by rapidly turning a light switch on and off (although this wouldn't make you very popular with the house owner). PWM works in a similar way, except that the on and off times are much smaller, and the switching is performed by a MOSFET. The HAT contains circuitry that generates PWM signals for four motors. All we need to do is call the Python **setMotor** function to tell this hardware to set the speed of a particular motor. The function accepts two parameters: the number of the motor to be controlled and a power level in the range of -100 (full-power reverse) to 100 (full-power forwards).

```
setMotor(1,50)
```

The above statement would ask **setMotor** to turn motor 1 forwards at half power. The motor driver code supplied by Hiwonder is shown in **Figure 1**. This function is in the **HiwonderSDK/Board.py** source file. The first few lines of the function (45–54) validate the incoming motor number and speed value. It is interesting to see how different errors are managed. An attempt to access a non-existent motor (for example, motor number 5) would cause an exception to be thrown and the function will interrupt the program that called it. However, an attempt to use speeds outside the speed range of -100 to 100 would result in the speed being capped at the limit values and the function continuing. In other words, trying to make the motor go at speed 200 would not generate an exception, the function would use speed 100. This can lead to a philosophical programming discussion of the type best held over a beer (Why not fail for all errors?), but we haven't got time for this because we want to make our ROS drive system.

The code from lines 56 to 65 creates an inter-integrated circuit (I2C or IIC) message and sends it to the motor controller to set the power level. If the first attempt to send the message fails, the exception handler is triggered and tries to send the message a second time. If the second attempt fails, the function will fail with an exception.

```
44    def setMotor(index, speed):
45        if index < 1 or index > 4:
46            raise AttributeError("Invalid motor num: %d"%index)
47        if index == 2 or index == 4:
48            speed = speed
49        else:
50            speed = -speed
51        index = index - 1
52        speed = 100 if speed > 100 else speed
53        speed = -100 if speed < -100 else speed
54        reg = __MOTOR_ADDR + index
55
56        with SMBus(__i2c) as bus:
57            try:
58                msg = i2c_msg.write(__i2c_addr, [reg, speed.to_bytes(1, 'little', signed=True)[0]
59                bus.i2c_rdwr(msg)
60                __motor_speed[index] = speed
61
62            except:
63                msg = i2c_msg.write(__i2c_addr, [reg, speed.to_bytes(1, 'little', signed=True)[0]
64                bus.i2c_rdwr(msg)
65                __motor_speed[index] = speed
66
67        return __motor_speed[index]
```

LET'S TWIST AGAIN

Now that we know how the robot motors are controlled, we need to connect these motors to ROS applications. In the previous article, we discovered that ROS nodes communicate using a 'publish and subscribe' model. We built a **button** node which published messages when buttons were pressed, and a **display** node which subscribed to messages and displayed them when they arrived. Our motor controller will subscribe to a topic and then drive the motors in response to messages published on that topic.

The Twist message in ROS can be used to express the direction we want to move the robot and the amount the robot should rotate when it moves. If we make the **motor** node understand Twist messages, it can be used with other robot applications. A Twist message contains two components: the linear velocity and the angular velocity. Both values are expressed as vectors with values for the X, Y, and Z components. This sounds much more complicated than the kind of 'ahead half speed' messages we'd like to use, but it does turn out to be a lot more flexible. →

Figure 1 ◆
If you've ever wondered what 'professional' code looks like, this is a good example

NICE PLACE TO WORK

ROS comes with graphical tools that you can use to build and simulate your robot application. Below, you can see the author's first attempt at designing a robot using the Unified Robotic Description Format (URDF). He used the RViz2 graphical tool and the joint state publisher to validate his robot description (which, in this case, needed some work). These tools require quite a powerful machine. The author used a Raspberry Pi 5 with a solid-state drive (SSD) for this part of the project. He then placed the code in a GitHub (**github.com**) repository and the smaller Raspberry Pi in the robot and loaded it from there.

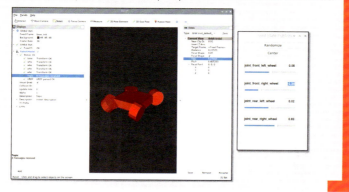

Above ◆
It turned out that the wheels were rotating on the wrong axes…

Figure 2 ◢
The Z value in this framework points straight up, i.e. out of the page

Figure 2 shows how the direction velocity values apply to the robot, which is pointing up the page. The X direction is forwards, the Y direction is left and right, and the Z direction is up and down. So, if I want the robot to move forward at one metre per second (which would be very fast), I'd set a linear velocity of (1, 0, 0). Angular velocity is the rate at which the robot should turn while it's moving. The rate is specified in radians per second. Angles can be expressed in degrees (with 360 degrees in a circle) or radians (with 2 * Pi – around 6.28 – radians in a circle). Working with radians is confusing, but it pays off when we start doing maths with the values. If both angular and linear velocities are both non-zero, the robot will move in a particular direction and turn at the same time. We

humans call this 'walking round a corner'. Whether or not a robot can move in a requested direction depends on the drive configuration. A car-like robot with steering wheels at the front will have difficulty rotating on the spot (i.e. it could not implement a Twist request with an angular component but no linear component). One of the great things about the mecanum wheel drive system is that it is possible for a robot to move in any direction and turn at the same time. We can use a bit of trigonometry to convert a Twist move request into the speeds each wheel should move:

```
def calculate_motor_speeds(twist):
    # Robot dimensions in meters
    L = 0.059  # Half the distance between front
and back wheels
    W = 0.067  # Half the distance between left
and right wheels

    # Convert twist velocities to motor speeds
    Vx = twist.linear.x
    Vy = twist.linear.y
    Vz = twist.angular.z

    # Calculations based on the mecanum wheel
formulae
    V_FL = Vx - Vy - (L + W) * Vz
    V_FR = Vx + Vy + (L + W) * Vz
    V_RL = Vx + Vy - (L + W) * Vz
    V_RR = Vx - Vy + (L + W) * Vz
    scale = 100.0 / max_wheel_power
    max_wheel_speed = max(abs(V_FL), abs(V_FR),
abs(V_RL), abs(V_RR), 1)
    setMotor(1,int(V_FL * scale))
    setMotor(2,int(V_FR * scale))
    setMotor(3,int(V_RL * scale))
    setMotor(4,int(V_RR * scale))
```

The `calculate_motor_speeds` function above takes in a Twist value and extracts from it the three values that are needed to control the mecanum motors. These are the linear X and Y values and the angular rotation. It uses a bit of maths, which the author doesn't quite understand (but seems to work), to calculate power values for each wheel (Front Left, Front Right, and so on). Then it scales the values so that they are in the range 0 to 100. Finally, it calls `setMotor` for each motor. Now that we have our motor power control sorted out, the next thing we need to do is make a ROS node that can receive Twist messages from other parts of the robot system.

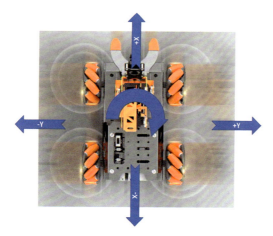

THE TOPIC OF MOVEMENT

The code below describes a `MecanumDriveSubscriber` object which will respond to messages to control the motor power. The `__init__` method runs when the object is created and creates a topic called `cmd_vel` which accepts Twist messages.

```
class MecanumDriveSubscriber(Node):
    def __init__(self):
        super().__init__('mecanum_drive_
subscriber')
        self.subscription = self.create_
subscription(
            Twist,
            'cmd_vel',  # Topic name
            self.listener_callback,
            10)
        setMotor(1,60)
        time.sleep(0.5)
        setMotor(1,0)
        self.get_logger().info("Motor subscriber
running")
```

The `__init__` method also makes a motor move briefly. This proved very useful when testing. It was obvious when the drive had been started. The class also contains a method called `listener_callback`.

```
def listener_callback(self, msg):
    calculate_motor_speeds(msg)
    return
```

The `listener_callback` method is called whenever a message is posted to the `cmd_vel` topic. The method just passes the incoming message to the `calculate_motor_speeds` function to move the robot.

TAKING CONTROL

The aim is to make a robot which will decide for itself which way it wants to move. But for testing, it is very useful to be able to control the robot directly from the keyboard. We can make a simple node which checks for key presses and publishes to a topic when it sees them. The code below creates a class called `KeyboardTeleop` which publishes movement instructions to the `cmd_vel` topic.

```
class KeyboardTeleop(Node):

    def __init__(self):
        super().__init__('keyboard_teleop')
        self.publisher_ = self.create_
```

```
publisher(Twist, 'cmd_vel', 10)
        self.timer = self.create_timer(0.1, self.
timer_callback)  # Poll at 10 Hz
        self.get_logger().info("Use WASD keys to
control the robot")
        # Save terminal settings
        self.settings = termios.tcgetattr(sys.
stdin)
```

The `__init__` method runs when the object is created. It does two things: it creates a publisher which can send Twist messages to `cmd_vel`; it also creates a timer which is triggered ten times a second. The timer will call a method called `timer_callback`. Let's have a look at the code for the timer callback. →

```
def timer_callback(self):
    key = self.get_key()
    if key == 'w':
        self.get_logger().info("w - forward")
        self.publish_twist(1.0, 0.0)  # Forward
    elif key == 's':
        self.get_logger().info("s - backward")
        self.publish_twist(-1.0, 0.0)  # Backward
    elif key == 'a':
        self.get_logger().info("a - left turn")
        self.publish_twist(0.0, 1.0)  # Left Turn
    elif key == 'd':
        self.get_logger().info("a - right turn")
        self.publish_twist(0.0, -1.0)  # Right
Turn
    elif key == ' ':
        self.get_logger().info("space - stop")
        self.publish_twist(0.0, 0.0)  # Stop
    elif key == '\x03':  # CTRL-C
        self.destroy_node()
        rclpy.shutdown()
```

DOCKER IS WONDERFUL

Docker (**docker.com**) is not a robotics tool. Docker is used to deploy large and complex applications. You create a 'Docker image' file which contains all the components you need for an application, and this image runs on your machine inside a software object called a 'Docker container'. The container sits on top of a layer which hides details of the underlying system. This makes it possible to deploy a ROS solution (which would otherwise need a particular version of Linux) on a standard Raspberry Pi. Learning a bit about Docker is a great idea; you can use it to deploy and manage all kinds of software.

The author has created a Docker image which contains ROS and a selection of useful tools that you can use with the examples in these articles. He has written a guide to setting up a Raspberry Pi, installing Docker, building a Docker image and running it, along with instructions for the exercises in this article. You can find it all here: **hsmag.cc/RosIntro**.

The function uses a method called `get_key` to get the currently pressed key. It then selects a movement option and calls a function called `publish_twist` for each movement.

```
def publish_twist(self, linear, angular):
    msg = Twist()
    msg.linear.x = linear
    msg.angular.z = angular
    self.publisher_.publish(msg)
```

The `publish_twist` method takes the linear and angular values and uses them to construct a Twist value (called `msg`) which is then published. This node can be used to drive the robot from the keyboard. If we start these two nodes running, we can use the keyboard to steer the robot around.

KEEPING IT REAL

We've seen that we can put different power values into motors to make them turn at different speeds. We've also found a way to take a movement request and convert it into speeds for each wheel. However, we have no way of mapping a requested speed onto a value to be sent to the motor controller. We know that putting a bigger number into the `setMotor` function will make the motor turn faster (and a value of 100 makes it run full speed), but we don't know how fast the motor moves. This works fine for remote control as the system above shows. It is quite fun to steer the robot around. However, we'd like to be able to make the robot move a particular distance, and at the moment this isn't possible.

We could start to solve this problem by obtaining the speed of the motors when our system turns them. For example, if we observe that a motor turns 60 times in one minute at full speed, that would mean the motor turns once a second. The wheel has a diameter of 65 mm. The circumference of the wheel would be Pi * 65 mm (around 190 mm). This means that at full speed, the wheel would move the robot 190 mm every second. Unfortunately, this would not work well in practice because the motor slows down when it is under load.

Figure 3 shows a motor which has a shaft encoder. The encoder sends pulses when it detects motor movement. We could use this signal to determine exactly how much a motor is turning and work out how far a wheel is moving the robot. Unfortunately, the motors in the Hiwonder robot don't have shaft encoders, so there is no way that the robot control software can know how far the motors have turned. The author decided to add a sensor to the robot to detect movement. This will provide feedback to a ROS application which can then adjust the movement commands. The sensor he has chosen is called an 'optical flow sensor'. It works in the same way as an optical computer mouse. The camera on the sensor detects changes in position of the surface beneath the robot and transmits X and Y values to a connected computer.

GETTING INTO THE FLOW

Figure 4 shows the selected sensor. The sensor has been mounted on a bracket which will be clipped underneath the robot. When the robot moves, the sensor will send movement information, giving the direction and speed of the movement it has detected. This information will be broadcast so that control software can monitor the robot's position.

ENTER THE PICO

The sensor uses an SPI connection to transmit movement values. There is no easy way to connect the sensor to the motor control HAT on the Raspberry Pi on the robot, so instead, a Pico was used to collect movement information and transmit it over a USB serial connection into the robot. The Pico was then plugged into a USB port on the Raspberry Pi. This worked well and means that the sensor can be easily added to any robot with USB connections.

Figure 4 shows how the sensor and the Pico are connected together. The following Python program was adapted from the example code for the sensor and runs inside the Pico to send movement information to the robot.

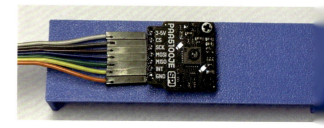

```
import time
import json
from machine import Pin
from breakout_paa5100 import BreakoutPAA5100 as
FlowSensor

flo = FlowSensor()
flo.set_rotation(FlowSensor.DEGREES_0)
led = Pin(25,Pin.OUT)

while True:
    delta = flo.get_motion(timeout=0.1)
    if delta is not None:
        x = delta[0]
        y = delta[1]
        result = { "x":x, "y":y }
        led.value(1)
        print(json.dumps(result))
    time.sleep(0.1)
    led.value(0)
```

This code runs inside the Pico (not on the robot). It connects to the sensor and then repeatedly reads it. If the sensor has new data, the values of x and y are assembled into a JSON message and printed to the console (which is the serial output for the Pico). A device connected to the sensor via the serial port receives the movement messages. Note that for this code to work, your Pico must be running the MicroPython image provided by Pimoroni. You can find it here: **hsmag.cc/PimoroniPicoUF2**.

ROBOT TRACKING

We now need a node which can receive the movement messages from the Pico and publish the values for a **controller** node to use. This will be a bit like the **keyboard** node we have just seen, except that rather than checking for key presses, the **timer_callback** method checks for messages from the movement sensor.

```
def timer_callback(self):
    if self.serial_port.in_waiting:
        data_str = self.serial_port.readline().
decode('utf-8').strip()
        try:
            data_json = json.loads(data_str)
            x = float(data_json.get('x', 0))  #
Convert to float
            y = float(data_json.get('y', 0))  #
Convert to float

            twist = Twist()
            twist.linear.x = x
```

POWER FOR TESTING

The robot is battery-powered and runs for around half an hour on the enclosed batteries. The author added a speed limiter to the `setMotor` function so that he could restrict the motor power and allow the robot to run on a mains adapter with less power capacity than the batteries. This proved very useful when developing the code.

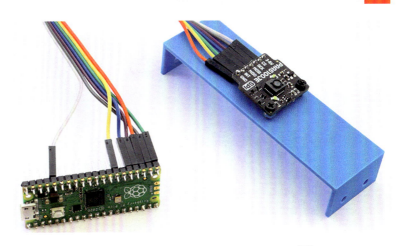

```
            twist.linear.y = y
            # Assuming z, angular x, y, and z are 0
or set them as needed
            self.publisher_.publish(twist)
            self.get_logger().info(f'Publishing:
{twist}')
        except json.JSONDecodeError:
            self.get_logger().error('Could not
decode JSON from serial data')
        except ValueError as e:
```

The code above shows the `timer_callback` function for the motion sensor node. It checks to see if there are any characters available from the serial port and reads a line if there is. Then it decodes the JSON and extracts the x and y values from it. These values are then used to create a Twist value which is published. A process wanting to track the robot movement can just subscribe to the events and be informed when the robot moves.

TAKING CONTROL

The author has created a drive package which contains the **keyboard**, **motor**, and **motion** nodes which can move the robot. He is presently working on a **controller** node which will use these to move the robot particular distances. He hopes that these articles have served as a useful introduction to the lower-level components of ROS. If you want to refine your robotics skills, you can work through the exercises on the project GitHub page, which you can find here: **hsmag.cc/RosIntro**. □

Above ◈
All of the connections are on one side of the Pico

QUICK TIP

When creating code that talks to hardware, it is useful to make something happen when the code is working. The sensor reader flashes the LED on the Pico each time it sends a movement message. This was very useful when debugging.

INSPIRATION

HACK | MAKE | BUILD | CREATE

A collection of quite brilliant projects to dazzle and inspire

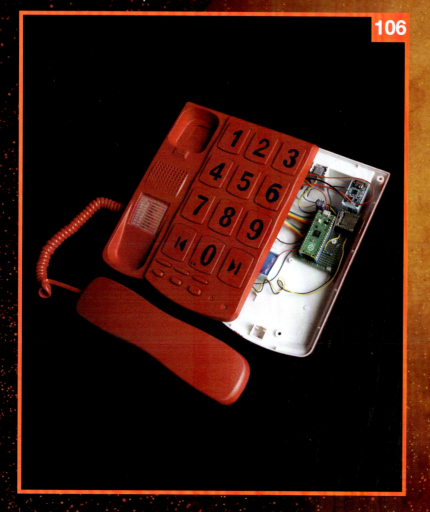

106

112

FEATURE

HOW I MADE

By Turi Scandurra

JUKEPHONE

Want some music? Just dial-a-tune

As an electronics tinkerer, I get a thrill from perusing the shelves of local charity shops in search of old gadgets I can creatively revamp. I look for things to take apart to see their inner workings, and uncover the system of mechanical and electronic parts that make them function. On one such quest, I came across a dismissed landline telephone with chunky keys, which was practically begging to be hacked into something else. The moment I saw it, ideas started swirling in my head about how I could give it a new life.

I initially focused on the keypad alone, thinking it could become a comically oversized numeric pad for a laptop (I might still do this). I would have then been left with a decent, spare speaker and a microphone. But I knew that with the right alterations, this retro relic could sing a new tune. So, in a musical twist, I came up with the idea of a personal jukebox, encased into the telephone, that would let me play tunes by dialling out their track number

I started by disassembling the device and inspecting it to figure out how to incorporate as many original components as possible into my new design.

What I used

> Landline telephone (I'm afraid a rotary dial phone won't do it)

> Raspberry Pi Pico

> DFPlayer Mini (or MP3-TF-16P clone) – MH2024K-24K, MH2024K-16SS and many more chips are supported

> microSD card – 8GB or more is recommended

> TP4056 battery charger module

> JST plugs

> 18650 or equivalent lithium battery

> 1000 µF electrolytic capacitor

> 2 × 1 kΩ resistor

> 3.5 mm audio socket

The keypad, with its playful large keys, was surely the main hardware feature and had to stay in its place. The little status LED and the piezoelectric buzzer, which I carefully desoldered from the PCB, seemed perfect to serve as feedback indicators. The telephone had a toggle switch on the back to set the ringer volume. I could have wired it to act as a power switch, but then I decided to lose it and use the spring-loaded switch underneath the handset instead. I believe this change was a clear steer towards a more accessible, human-centric design. There is some beauty in the way you answered a call on a landline telephone – you would just lift the handset and listen. With no buttons to press, it was just the natural gesture of bringing the speaker to your ear. That's the right level of simplicity that I wanted for the final users of my new object. And after all, listening to music through the handset speaker is what gives this music player its quirky character.

Since the original PCB was inevitably going to become e-waste, my creation needed a new brain. The microcontroller I use most often these days is the RP2040, specifically in the form of a Raspberry Pi Pico. It's powerful but also energy-efficient, offers plenty of GPIO pins, and its price is very affordable.

Above ◆
Raspberry Pi Pico provides a powerful and affordable brain

FEATURE

Above ↑
The phone's battery is recharged via USB-C

" I USED FOUR C LIBRARIES"

While the Pico can be programmed to produce audio output via pulse-width modulation or I2S – like I did with some of my previous projects – its capabilities are not quite right to get high-quality music playback. So, I picked a separate music player module, the popular DFPlayer Mini. It can play audio files from a microSD card, has a built-in amplifier to drive a small speaker, and can be controlled digitally via UART communication.

Armed with my soldering iron, I first rewired the telephone's keypad through a ribbon cable and to a perfboard on which I had soldered two pin header sockets for the Pico. The connection between the keypad and the Pico GPIO pins allowed me to start writing code and let the controller detect key presses.

For this project, I used four C libraries that I had already written and one created by another developer that I ported to the Pico. I like to break down my code as much as

possible into a modular structure that lets me reuse components between projects, and every project is an opportunity to write new libraries.

C LIBRARIES

The most recent one is RP2040-DFPlayer, which implements the UART communication protocol to send instructions to the MP3 player and poll its status. Without it, the features of the Jukephone would have been limited to only basic actions like starting the playback of the first track, dialling the volume up or down, and skipping to the next track. I took time to study the datasheet of the player and unlocked features like equaliser presets, playback modes, and querying of playback status.

RP2040-Keypad-Matrix is another crucial component of this project's software. It alternates write/read cycles across the rows and columns of the matrix to poll it for changes, and discern between short and long key presses.

Even though the buttons produce a tactile feedback on their own, I made it so that the little LED blinks shortly after each key is pressed, together with a short beep emitted by the phone buzzer. For that, I used RP2040-PWM-Tone, my tone generation and melody player library for Raspberry Pi Pico. I initially used it to play a start-up melody, but finally settled for a quiet start-up because I feared a jingle would get annoying in the long run. When designing products, it's tempting to add features just because the tech allows you to, but 'possible' does not mean 'necessary'. Ultimately, this is an open-source project, so adding the melody back would require just one line of code, since a few sample melodies come bundled with the library.

A nice little utility that I add to all my machines is provided by the library RP2040-Battery-Check, which, as the name suggests, periodically checks the battery voltage and rapidly flashes the LED when it's time to recharge it.

The last library I included is RP2040-Button (the one I ported from another developer's work), which is needed to detect presses of one lone key that is not wired with the rest of the matrix.

The main software logic is pretty straightforward. Key presses are debounced, numbers concatenated and clamped between 1 and 999. When a valid track number is entered, after a short timeout the MP3 player is sent a command with the ID of the new track to play. The chip on the MP3 player is able to pick a random track to play on its own, but I added my own playlist randomisation function on the Pico, which means that tracks do not repeat until the whole playlist has played entirely. However, the Jukephone I set up contains over two days' worth of music, so I don't think it will happen often.

UART communication between the Pico and DFPlayer uses just two wires for TX and RX, one of which is filtered with a 1 kΩ resistor to reduce noise.

POWER PARTICULARS

The output pins of the MP3 player go straight to the speaker inside the handset. There's also a 3.5 mm mini-jack socket, connected to the player's built-in DAC, so you could plug headphones or external speakers into the back of the Jukephone. Sound quality is much higher through this output and, unlike the handset, it's a stereo output.

The whole project is powered by a 3.7 V lithium battery, recharged by a TP4056 module via USB-C. This little module is a staple for all my portable designs as it also provides overcharging and undercharging protection, for prolonged battery life. I can say I treat my batteries fairly well because I never heard them complaining. →

Below ↓
I'm pleased with the finish on the red paint

JUKE

Above ◆
**Perfboard is great
for one-off projects**

" I HID AN EASTER EGG IN MY BUILD"

It might have been overkill here, but I've taken the habit to always add a decoupling capacitor of around 1000 μF just after the power stage to help stabilize the voltage and squash any ripples that might pop into my circuits. It's worth noting that technically the charging module could simultaneously provide energy to the battery and the rest of the circuit, but doing so is not recommended as it interferes with the charge cut-off, increasing the risk of overcharge.

Opting for no changes to the original plastic housing meant that I had to repurpose the openings for two toggle switches and for the RJ11 socket that once connected the telephone to the wall. They became natural slots to access the microSD card, the USB-C charger, and the headphone jack.

FINAL ASSEMBLY

My perfboard – holding the Pico, the player, and the few discrete components – was perforated further in order to open two mounting holes and secure it to the housing.

I 3D-printed a custom mount for the TP4056, which locks nicely around the housing wall and a standoff, while also raising the module to its perfect placement and angle. The lithium battery, connected to the charger via a pair of JST plugs to make it swappable, was secured to the housing floor with a dab of hot glue.

The telephone's exterior went under a DIY makeover with a few coats of spray paint. And no other colour screams 'touch me' like shiny red! I then labelled the keypad buttons with metallic purple paint markers. The coiled handset cord is the only part I needed to replace, as painting over the old one was not going to offer a durable finish.

I loaded the microSD card with 999 MP3 files, organised so that there's one hundred per genre (except the first 99). Specific tracks can be invoked by typing their number on the keypad. I replaced the old directory card under the handset with a printout of the genres available.

PHONE

The six additional keys below the large numeric keypad were programmed to perform useful actions, like increasing and decreasing the volume, pausing and resuming playback, restarting the current track, toggling repeat mode for the current track, playing a random track, and rotating between five sound equalisation presets.

In thinking of a future iteration of this project, my mind goes to the possibilities of an internet-enabled device. While still retaining the central feature of track selection by numeric input, it could enable a Wi-Fi back-end to customise the playlist with web resources to stream.

I hid an Easter egg in my build. Dialling a specific number triggers a voice message

that says: 'Thank you for calling the Jukephone Helpline. All our operators are busy at the moment, please try again later.'

I can imagine the small person who will receive my first Jukephone as a gift getting ready for bed, tapping out children's tunes until she gets sleepy. Hopefully it will stay functional until she's old enough to appreciate the rest of the playlist.

Finding an old gadget a new home rather than the landfill is rewarding both creatively and environmentally. I can feel my Jukephone evoking a sense of nostalgia for relics of the past, but in the end it represents the joy of second chances. And even though it's not really a telephone any more, it still does phone home in a certain way. □

LINKS

> turiscandurra.com/circuits
> github.com/TuriSc/Jukephone

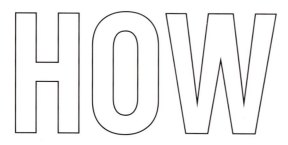

By **Karl Mose**

PIARTFRAME

Bringing mathematical art to halls of residence

About four years ago, in the cold of December, I was boarding a plane destined for the UK. I'd applied for university to study computer science, and part of the interview involved an expectation you'd have some kind of project to talk about. So, I spent the preceding weeks attempting to devise a project that I hoped would impress the interviewers. I'd always been very fascinated with maths and, at the time, fractals more than anything had caught my attention. Fractals are geometric shapes that we can generate with mathematics. Some fractals, when tuned correctly, can be stunningly beautiful. To this day, it still blows my mind how much visual complexity you can have emerge from relatively simple equations. At that time, I spent hours exploring their vast landscapes and adjusting their colours, making it a natural choice for the project.

The project that I came up with at the time was a rather painstaking effort of a couple of months, using a Raspberry Pi to generate the fractals, and a smartphone app that would download them periodically and set them as my phone wallpaper, for me to look at throughout the day. The app

FEATURE

"WE FIRST PICK A SET OF BOUNDARIES"

some decoration. And so, I decided to pay a homage to this old project of mine that originally led me to this university path. That's how I came up with the idea for the PiArtFrame: rather than rendering fractals to my mobile phone, I'd turn them into a piece of room decoration. I wanted to build a digital picture frame that, using the Raspberry Pi, would continually update itself with new fractals.

The first hurdle of the project would be to write code to actually render the fractal. Speed wasn't essential, as the frame just needed to run fast enough to update a couple of times a day, so I opted for Python. The goal of the script was to start at some large-scale section of the Mandelbrot set, and then continually zoom in on increasingly smaller areas that appeared visually interesting. To render a section of the Mandelbrot set, we first pick a set of boundaries. For example, this could range from x = -2 to x = 2 and y = -1 to y = 1. We then divide this section into a grid of pixels and, for each pixel, we compute whether or not it is contained in the Mandelbrot set, using the aforementioned algorithm. For those who

would like a more in-depth look at how one computes whether or not a point is in the Mandelbrot set, check out the GitHub repo of the project (**hsmag.cc/PiArtFrame**).

A notable challenge arises from the script continuously zooming into the Mandelbrot set, necessitating increasingly precise floating-point computations. To solve this, I used the Python 3 decimal library, which allows for doing computations at arbitrary precision. As the renderer zooms into a smaller and smaller section of the Mandelbrot set, it increases the precision at which it does decimal computation. As one attempts to render increasingly granular sections of the Mandelbrot set, it takes longer and longer to determine whether or not a given point is in the set or not. The algorithm we use needs progressively more iterations, and to make matters worse, there's not an easy way to calculate *how* many we need. If we do too few, the render will look smooth

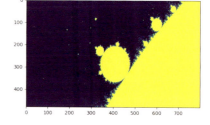

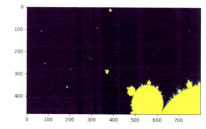

and uninteresting, and if we do too many, it will take too long. I made an approximate equation for the number of iterations for a given zoom-level, but it is inexact and sometimes either over- or undershoots.

One of the hardest parts of the script was designing a method for automatically exploring the Mandelbrot set. The Mandelbrot set is huge, but most parts of it aren't that visually interesting, it's only really around the edge that you see the interesting patterns discussed earlier. The current system that handles exploration works by dividing the most recent render into four squares. It then quantifies the uniformness; that is, how close each of the four quadrants is to being just one colour. Based on this info, it then picks a random quadrant to zoom into, filtering →

Right →
Fractals occur all over the place in nature, most deliciously in the Romanesco – a sort of halfway house between broccoli and cauliflower

Credit: Ivar Leidus
creativecommons.org/
licenses/by-sa/4.0/

Below ↓
All the parts you need: the frame, picture mount, back panel, e-ink panel, and the Raspberry Pi Zero

WHAT ARE FRACTALS?

For those who've yet to be initiated, consider this a much too brief introduction into a very exciting and mind-boggling world. Fractals are loosely defined as geometric objects that display detailed structure at arbitrarily small scales (source: Wikipedia). That means that one can keep zooming in on a fractal, without it ever reducing down to a pixellated grid like a traditional image, or just lines and colours like a vector graphic. Many fractals are self-similar – that is, the fractal contains itself in its entirety within some subsection. Fractals sometimes appear in nature, in things like pine cones, snowflakes, or even broccoli! The Mandelbrot is just one example of a fractal, but a rather interesting one.

The Mandelbrot set is defined by an equation which tells us, given a certain two-dimensional coordinate, whether or not that point is inside the Mandelbrot set or not. It's based on a relatively simple equation using complex numbers. If you're familiar with complex numbers, it won't take more than two to three minutes to understand the algorithm by which we figure out if a pixel is included or not included in the Mandelbrot set. There are plenty of good sources online, but I can recommend the YouTube channel Numberphile in particular. If you've not yet heard of complex numbers, I'd highly recommend it – they allow you to take the square root of a negative number, which is quite cool. The Mandelbrot set is self-similar – but often in surprising ways. The Mandelbrot set contains itself an infinite amount of times, but in some cases, it may also be bent or skewed in some odd way! There's loads of other different shapes to discover within the set, with some areas having been dubbed with creative names such as the Sea Horse or Elephant Valleys.

FEATURE ━━━━━━━━━━━━━━━━━━━

out any fully uniform quadrants. This algorithm isn't perfect, and sometimes (though not often) leads one into fairly uninteresting areas of the Mandelbrot set. If one sees that the frame has ended up in an uninteresting section, though, all it takes is a quick switch on and off the Raspberry Pi to restart its search. The script can take anywhere from a few minutes to hours or days to render, depending on how far one has zoomed in. This could be sped up (to a limited extent) using a faster Raspberry Pi. That said, part of the charm of the project is that the frame updates only rarely, and considering the superior power characteristics of the Raspberry Pi Zero, it seemed a natural choice for this project.

The project's aim was not to create something large and attention-demanding, but rather something subtle that could be mistaken for an ordinary picture, if not for the power cord. The natural choice for such a project is an e-ink screen. I picked a 7.5-inch e-ink screen from Waveshare, which is sold with a Raspberry Pi header that makes integrating with the Raspberry Pi Zero a breeze. While the e-ink does add to the subtlety, standard e-ink panels still have a grey hue rather than white that distinguishes them from, well, paper. Leaning on a past summer job working in a print shop, as well as advice from friends with better taste than myself, I decided to try to give the frame a more retro look to mask the tech-vibes of the e-ink colours. I paired the screen with a comparatively large 18 × 24 cm light-wood picture frame which, when paired together with a white picture mount, would give the frame a more classic look. Other than aesthetics, the picture mount also provides padding between the glass of the frame and the e-ink panel, which makes me more confident in the durability of the build. I also configured Linux to start the script when the Raspberry Pi's operating system boots.

The build itself is very simple, and can be done with the aforementioned pieces

Left ←
Only minor modifications to the picture frame were necessary

"THE BUILD ITSELF IS VERY SIMPLE"

of hardware, as well as a knife (ideally a penknife, but anything sharp will do). Simply insert the picture mount inside the picture frame, and then place the e-ink panel on top. Make sure that it's aligned – if you'd like, you can use tape to keep it in place, but it's not strictly necessary. You'll need to cut a little slit in the back of the cardboard for the e-ink panel's cable to go through. Most picture frame backs are made of cardboard or some kind of fibre, so this shouldn't be too difficult. The slit doesn't need to be large, but ensure that the angle on the frame isn't too sharp. Now all you need to do is insert the back of the picture frame, and squeeze the tabs tightly so that the panel doesn't slip. The mount provides a cushioning for the panel here – if you're not using a mount, I'd recommend taping the panel instead and only pushing in the tabs lightly.

The project is done, for now. In the future, I'd like to improve the exploration and iteration-approximation algorithms further so you're more guaranteed to see an interesting image. I'd also like to experiment with the new colour e-ink panels, as the Mandelbrot set can look stunning with some simple colouring.

The last challenge was picking a spot for the finished frame. I couldn't decide between the office or my bedroom, but chose the latter. Now, I wake up to new artwork every day, and it's quite exciting when I catch it changing. It reminds me of the intrinsic fun factor in computer science, and why I decided to pursue this degree to begin with. □

Above ↑
The frame, as well as the Raspberry Pi Zero, now have a home on my pinboard

Above Left ↖
Good thing that this side is facing the wall

HOW I MADE

By **AAED MUSA**

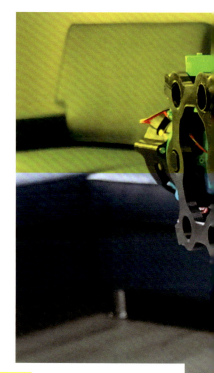

TOPS - TRAVERSER OF PLANAR SURFACES

Quadrupedal robots are quite peculiar. While most robots perform tasks that humans and animals can't quite accomplish, quadrupeds mimic natural movements, which is harder to do.

This very challenge is what has inspired me to build several of my own robot dogs over the years. So, what exactly goes into building a robot dog? Just how close can you come to replicating an actual dog? And how cheaply can this be done?

Setting out to answer these questions led me to build my latest quadruped named TOPS (Traverser of Planar Surfaces). TOPS is an open-source twelve degrees of freedom (DOF) quadrupedal robot that can walk, trot, and dance. Other than the motors, bearings, and screws, TOPS is fully 3D-printed.

The name TOPS also spells SPOT backward: SPOT being the famous Boston Dynamics robot.

This project was heavily inspired by the YouTuber James Bruton, who built openDog V3, another open-source quadrupedal robot. The goal of this project was not only to create a functional robot, but to make a robot that provided a sense of realism. Getting a quadruped to walk is one thing, but getting a quadruped to walk with dynamic motion is another.

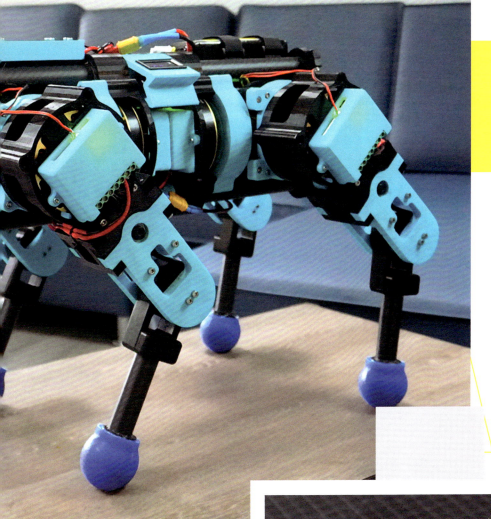

MAKING MOTION

The actuator design of TOPS was the first step in the build process. The most ideal actuator design for a quadrupedal robot is one that follows the quasi-direct drive (QDD) scheme. A QDD actuator can be defined as an actuator that has a low enough gear reduction (under 10:1) to retain the benefits of a direct drive actuator (like efficiency, speed, and backdrivability) while also having a high torque output. To ensure that high torque is achieved with a low gear ratio, it is best to use a high-torque motor. I decided to use Eaglepower 8308 90KV brushless motors that I found on AliExpress for only $60 each. Apart from being cheap, I chose these agricultural drone motors because of their flat and wide pancake-like shape.

Pancake-style brushless direct current (BLDC) motors have high torque densities due to having a large gap radius or distance from the centre of the motor to the rotor. I found the theory and design of an optimal actuator best summarised in an article by the MIT Biomimetic Robotics Lab entitled 'Optimal Actuator Design' (**hsmag.cc/OAD**).

Above ↑
These motors gave
me plenty of power
for just $60

After motor selection, I had to figure out the gear reduction. Planetary gear drives are the common choice for QDD actuators due to their compactness and simplicity. I decided to go with a 9:1 planetary gear reduction since it puts the peak theoretical speed of the actuator at around 222 rpm at 22.2 V. This seemed to be more than fast enough. I used helical gears since they provide gradual contact between engaged teeth, making them both quieter and able to withstand higher loads when compared to standard spur gears.

Motor control was the next item on the actuator design list. Motor controllers are what transform a purely mechanical actuator into a dynamic robotic limb. While drones and RC cars use ESCs (electronic speed controllers) to control a brushless motor, field-oriented control (FOC)

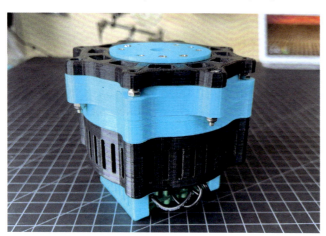

controllers are the preferred boards for robotic applications.

FOC is a control method for brushless motors that allows for closed-loop position, velocity, and torque control. FOC controllers can smoothly drive a BLDC motor with an attached load, which is why they are best suited for robotics. FOC controllers essentially turn a BLDC motor into a virtual spring that can be dampened or stiffened by changing different gain parameters. This added compliance to a robot's actuators enables abilities like shock absorption, recovery from being pushed, and the ability to walk across uneven terrains. The FOC controller that I chose to use for this project was the ODrive S1. The controller setup is quite simple: the motor's three phases connect to the controller and an encoder magnet is added to the motor's shaft. An onboard or external encoder is then able to read the motor's position. From there, the motor can be calibrated and configured with different settings.

ALL TOGETHER

The full actuator design has three sections: bottom, middle, and top. The bottom section houses the ODrive controller which reads the position of the brushless motor using the ODrive's onboard encoder.

The middle section houses the brushless motor. This section features slots around the radius of the housing which act as air vents for the passive cooling of the motor.

The top section houses the planetary gear-box and the output shaft. The sun gear is directly mounted onto the BLDC motor and the planet gears are suspended on a planet carrier. The actuator utilises

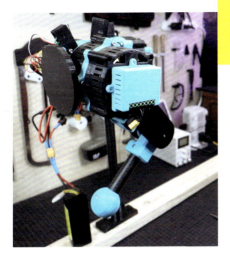

Left ←
The first leg design was a bit big and heavy

13 3D-printed parts that I printed on my Creality CP-01.

Below are the specs for a single actuator:

- Total costs: $247
- Total weight: 935g
- Dimensions: 133 mm × 105 mm
- Peak torque: 16.36 Nm.

FIRST LEG DESIGN

With the actuator design complete, the next step in making TOPS was to design a single 3DOF leg. I went through two different leg designs before moving to the full robot design.

A 12DOF quadrupedal robot is made up of four legs, each leg having three actuators. The three actuators are the abduction/adduction (abad), knee, and hip actuators; each is named after the joint that they control.

One design choice that was made early on was to integrate parts of the leg design into each actuator's design rather than designing the leg around a standard actuator. In other words, each of the actuators has a slightly different exterior design. This helped to limit the number of extraneous parts needed.

The knee actuator, which rotates the forearm, uses a 1:1 belt pulley reduction to reach the knee joint. The forearm is made from a carbon fibre tube. The foot is in the shape of a sphere and was simply 3D-printed for this prototype. In total, the single-leg prototype weighed 3.42kg.

To test the leg, the actuators were connected to a Teensy 4.1 microcontroller via UART.

While the ODrive's onboard encoder can measure the absolute position of the BLDC,

it cannot measure the absolute position of the output shaft due to the 9:1 reduction. I therefore decided to use limit switches to home the actuator.

The first step in programming the leg was to derive inverse kinematic equations to accurately place the foot in a known position in 3D space. The derivation process largely consisted of simplifying the leg design into a series of lines and solving for different angles by forming triangles with those lines. These equations allow the computer to know exactly how to position each actuator to place the foot at an input X, Y, Z position.

SECOND LEG DESIGN

Following the first leg design, I saw room for improvement in the areas of weight, communication, homing, and overall aesthetic.

To reduce the weight of the leg, I decided to add weight-reducing slots and holes to all of the parts; however, the biggest weight reduction came from completely redesigning the knee actuator. Previously, the knee actuator design, like the other actuators, had a 9:1 planetary gear-box. It was then connected to a 1:1 belt pulley system to reach the knee-joint. This time, I decided to remove the planetary gear set from the knee actuator and incorporate the 9:1 reduction into the belt pulley system instead. This also reduced the overall width of the leg, therefore needing less torque to be applied to the abad actuator.

The second leg design prototype weighed 2.98kg, which is 0.44kg lighter than the initial prototype.

Below ↓
The actuator brings together the motor and the gear-box

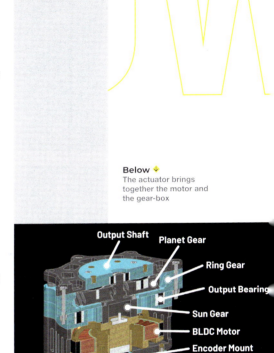

- Output Shaft
- Planet Gear
- Ring Gear
- Output Bearing
- Sun Gear
- BLDC Motor
- Encoder Mount
- Air Vents
- ODrive S1

Left ←
All the actuators for one leg

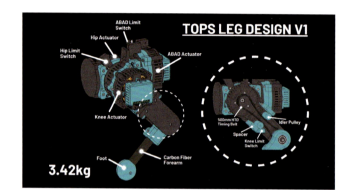

TOPS LEG DESIGN V1

ABAD Limit Switch
Hip Actuator
Hip Limit Switch
ABAD Actuator
Knee Actuator
500mm HTD Timing Belt
Idler Pulley
Spacer
Knee Limit Switch
Foot
Carbon Fiber Forearm

3.42kg

Right →
Leg design V1 drawing

Above ↑
Leg design V2

Previously the communication protocol used to communicate the ODrives with the Teensy was UART. Unfortunately, the Teensy only has eight UART ports when twelve are needed for the full robot. I decided to switch to CAN bus communication.

CAN bus is commonly used in modern vehicles, and it simplified the wiring as only two pins on the Teensy are needed: CAN High and CAN Low.

Another change made in the second prototype was to get rid of the limit switches for homing and, instead, utilise the physical limits of each joint. When performing the homing sequence, each joint is moved to its physical limit. Since this position is unchanging, it can be considered the home position. This home position can then be worked into the inverse kinematic equations as constant offset values.

FULL ROBOT DESIGN

With the second prototype made, I felt comfortable moving onto the full quadruped design. I used four carbon fibre tubes to make up the frame of the robot. Each leg then slides onto two of the tubes and is secured with clamps built into the abad actuator housing. The front of the robot houses a small 16×2 LCD screen to show the robot's current mode of operation.

TOPS is controlled with an eight-channel RC remote and is powered by a 6S (22.2 V) 5200mAh LiPo battery. The actuators are directly powered by the battery while the Teensy is powered by a 5V regulator. The battery is monitored by a voltage display on the left side of the robot.

It took three weeks to print everything on my single printer, after which I had to put everything together.

First, I built the twelve actuators, then the four legs. Finally, I brought these together to make the full robot.

I decided to cast the feet in 30A silicone, which is somewhere between the squishiness of a rubber band and an eraser. First, I had to 3D-print the sole and the mould.

To cast the feet, each sole was suspended in the mould and then I poured in the A-B silicone mix to cover the surface of the sole. As a result, the foot has a squishy and high-traction silicone exterior and a rigid 3D-printed sole that allows it to be connected to the forearm.

GAIT SEQUENCING

The basic principle of programming a quadruped to walk is to have a pair of diagonal feet in contact with the ground at any given moment: this maintains balance and is otherwise known as trotting. Getting a single foot to take a step forward involves four commands:

1. Move foot up
2. Move foot forward
3. Move foot down
4. Move foot backward to original position

Getting the robot to walk forward is essentially a two-step sequence. First, two diagonal feet must simultaneously take a step. As these feet touch the ground, the two other diagonal feet begin to take their

Below ↓
Assembled QDD actuators

step. As those feet begin to touch the ground, the sequence is repeated.

To walk in another direction, step 2 of the above step sequence is replaced with the desired direction of travel (backward, right, or left), and step 4 is replaced with the direction opposite to the direction of travel (forward, left, or right) to move the feet back to their original positions.

To rotate the robot, a slightly different approach is taken. In this case, diagonal feet are moved sideways but in opposite directions. While the sequences themselves are simple, programming them is a bit more challenging. There are many variables to consider, like the amount of time each leg is off the ground, how far the legs lift off the ground, how big each step is, and how fast the legs accelerate. These factors not only determine if the robot can walk in the first place, but how dynamic that walking motion is.

LOOKING BACK

- Total cost: $3300
- Total project timeline: three months
- Total weight: 13.4kg
- Weight of 3D-printed parts: 4.5kg
- Full battery run time: 10–15 mins

Constructing a robotic system from the ground up was a great engineering exercise. As it turns out, a 3D-printed planetary gear-box can work quite well for large-scale projects with heavy loads. I plan to continue to try out different gearing, motors, and manufacturing methods for more optimal actuator designs. In terms of dynamic gait, I think that TOPS performed

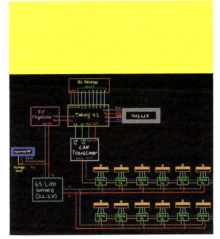

Left ←
TOPS circuit schematic

fairly well; however, there is lots of room for improvements.

The main mechanical limitation of this project is that the knee-joints skip a lot due to the small contact area between the belt and the smaller gear. Tensioning the belts did not fully solve this issue. In the future, I think it would be best to keep a similar configuration to the initial leg design and have the gear reductions built into the knee and then use a 1:1 belt pulley.

One of the shortcomings of this project on the software side is having to manually calibrate each leg on startup. This became quite a labour-intensive process throughout the testing phase.

Another software shortcoming was a lack of environmental feedback. In the future, it would be best to use an inertial measurement unit (IMU) so the robot can correct itself if it encounters something unexpected.

I also feel that the stepping sequences are a bit rudimentary and could be smoothened to reflect a more natural gait. I hope to address these in my future robot dog projects. Until then, it's safe to say that TOPS is my most advanced robot dog to date. □

Below ↓
TOPS dancing

Below ↓
Leg Design V2 Drawing

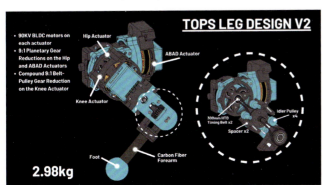

TOPS LEG DESIGN V2

- 90KV BLDC motors on each actuator
- 9:1 Planetary Gear Reductions on the Hip and ABAD Actuators
- Compound 9:1 Belt-Pulley Gear Reduction on the Knee Actuator

Hip Actuator
ABAD Actuator
Knee Actuator
Foot
Carbon Fiber Forearm
300mm HTD Timing Belt x2
Spacer x2
Idler Pulley x4

2.98kg

HOW WE MADE

By **STÉPHANE GODIN**

AN OPEN-SOURCE BRAILLE EMBOSSER

Imagined in 1829 by Louis Braille, **Braille is a tactile alphabet for blind and partially sighted people.** Made of a grid of six dots, each Braille cell corresponds to a letter or an escape code to specify a capital, number, or punctuation mark. As each cell consists of six dots, there are 64 combinations available. Due to the limited number of combinations, each country has adopted its own Braille standard. The usage of computers and information technology shows the need for a better standard with more combinations. Since the start of the 21st century, we also have eight-dot Braille standards. At this time, eight-dot standards are used only on numerical devices, such as Braille displays.

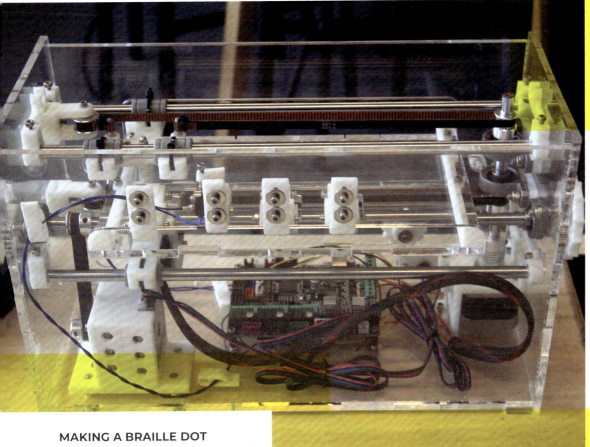

MAKING A BRAILLE DOT

When creating Braille documents, a simple
method is to use a metal needle and an
elastic material, such as a mouse pad,
placed behind the paper. You create the
Braille dots by pressing the needle gently
onto the paper, causing it to deform.
However, the dot will have a blurred edge.
We started this project by using an anvil,
and found that using a 2 mm hole in a flat
3D-printed part produced better results.
This method created a dot with a nice
relief over the paper and a sharp edge.
Unfortunately, the 3D-printed parts proved
to be unreliable, as the material fell apart
after a few dots. After testing, we found
that the anvil must be made of metal. We
also found that the hex hole of an M3 grub
screw is perfect for our needs. As a result,
all of our designs are now made with a
30 mm grub screw. The hex hole serves
as the anvil shape, and we glue and lock
a hex nut onto the other end to adjust the
depth of the Braille dot.

PLANNED DESIGN

Our plan was to create a standard 2D board
with a Braille needle on a carriage, building
on previous tests conducted on modified
CNC or 3D printers. Since Braille dots are
always at the same depth, we don't require
precise control of the tool depth. Therefore,
a simple solenoid is sufficient to manage
the up and down positions. A solenoid is
much faster than a stepper motor or servo.
Just think about a standard CNC with a
Braille needle on a solenoid as a tool. As
we've seen, a good Braille dot requires a
needle and an anvil. Therefore, our Braille
tool consists of two parts. The paper is
placed between two carriages, with the
bottom one using a solenoid to punch dots
while the other acts as the anvil. The paper
is fed with stepper-controlled rollers to
control the position. →

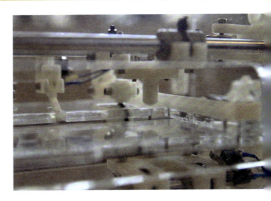

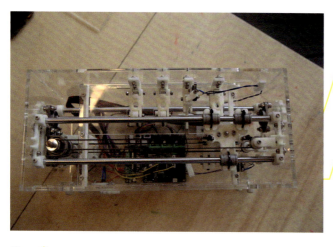

PAPER FEED

When handling the paper, the sheet must
be securely held between the two rollers.
The bottom roller is driven by the Y motor,
while the top roller applies the required
pressure. If the sheet slips on the rollers,
it can lead to inaccurate printing or even
cause a paper jam.

We created our first prototype using
three rollers and paper pressers. The
pressers were all situated on the same
8 mm linear rod axis, with two springs
at each end to ensure good pressure on
the paper. Although this system worked
adequately, it was difficult to build and
adjust. So, we decided to experiment with
three independent pressers and relied
on the plastic's elasticity to maintain the
pressure. However, the paper feeding
system was still tricky to adjust and was
not very reliable.

We then decided to try a simpler system
using only two rollers, as it is easier to align
two points. We had what we thought was
a good design that was simple to build
and reliable. We made a few prototypes
using this design and had some success.
However, with only two rollers, you can't
use materials like aluminium cans.

We then attempted to upgrade the paper
presser by adding springs, one for each
presser. We found the GT2 tensioner spring
was both affordable and easy to find. So,
we redesigned the presser with two plastic
parts articulated around an M3 screw, using
a GT2 tensioner as a spring. Success! This
new presser is more flexible and applies
firmer pressure on the paper, resulting in
a better paper roll that securely holds the
sheets of paper.

DETECTING THE EDGE

On our board, we need a reference to
define an absolute X and Y position on the
paper. On the X axis, there is a standard
end switch that detects the minimum
position of the bottom carriage and defines
it as the 0 position for X. The Y axis runs
along the paper sheet, so we must detect
the start of the paper. Initially, we used a
standard end switch with the end switch
lever in contact with the paper to detect the
top edge of the sheet, defining the top edge
as the 0 position for Y. It worked fine, but it
was difficult to set up – the paper jammed if
it was too close to the end switch and was
often undetected if too far. To improve

"WE HAD WHAT WE THOUGHT WAS A GOOD DESIGN THAT WAS SIMPLE TO BUILD AND RELIABLE"

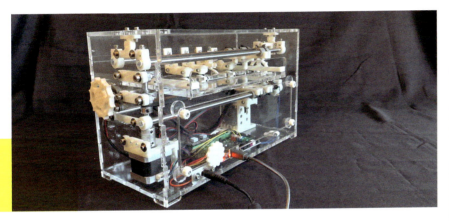

Left ←
The complete machine ready to run

Below ↓
Maybe I'm biased, but I think this looks good as well as being functional

the design, we used a 3D-printed lever to offset the end switch and a hex screw for fine-tuning the detection position.

PAPER PATH

Another important aspect of the paper feeding process is the path that the material takes within the device. It's a delicate process, and even the slightest obstacle can cause a paper jam. Initially, we used a plywood or PMMA plate with a 10 mm path for the Braille needle in the centre. However, this setup often resulted in paper jams. As the paper travelled downwards, sometimes one of its corners got caught on the edge of the needle path, causing a jam. To overcome this problem, we experimented with various 3D-printed designs to keep the paper securely on the tray. However, these designs required post-processing, and we found that it's better to let a piece of paper go its own way if it's sticking in the printer. We can lower the tray on the output side, just after the needle's path, and it works incredibly well! The lesson we learned is to never force a piece of paper and instead let it travel through freely.

SOLENOID ALIGNMENT

One recurring issue we encountered was the Braille needle would fail to lower when the solenoid was activated and deactivated, resulting in paper jams or missing Braille dots. The bottom carriage proved the most challenging part to create. The solenoid axis needs to move freely, so we designed a 3D-printed part that acted as a guide for the needle. Though everything seemed to work fine when we tested the Braille needle manually, we found that the needle would often lock in the top position when we used the electromagnet with power. As we were using a cheap electromagnet, we noticed that the position of tapped holes on the side of the solenoid was not accurately aligned with the solenoid axis. In some solenoids, there is a slight angle between the axis and the anvil. This can cause a lateral force on the axis when the needle enters the anvil, which is enough to lock the solenoid in the activated position. We found that the best solution was to make oblong holes for the solenoid fixing screws – this allowed us to finely adjust the solenoid axis with the anvil. →

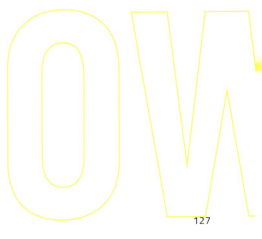

FEATURE

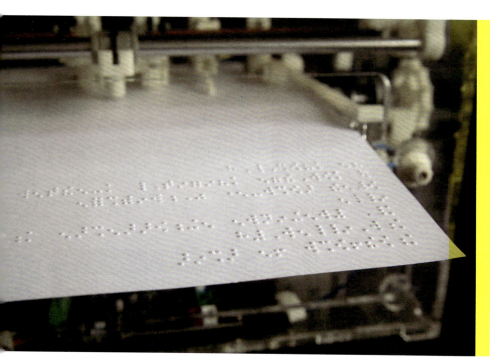

A WORD ABOUT SOFTWARE

We now have a device that enables us to move a Braille needle along the X axis using a stepper motor, and we can move the paper along the Y axis using another stepper motor. This is perfect for a 3D printer controller board with Marlin firmware. We modified the Marlin firmware to slightly adjust the Y homing. If the end switch is on, detecting the paper, we move the paper backward until the switch is off. If the end switch is off, we move the paper forward until the switch is on. This special homing feature allows the user to easily load the paper into the rollers, and the 0 Y position is automatically set at the top edge of the sheet.

The power output designed for the heatbed is used for the solenoid control. Initially, we used the G-code M3 command to enable and disable the solenoid. This command is typically used for laser control, where M3 S1 activates the laser, and M3 S0 deactivates it. We used the command in the same way to control the solenoid. However, solenoids

Left ←
The dots left by the machine are easy for people to feel with their fingers

are different from lasers; you can't activate them for too long. Without a cooling system, you must respect an activation/deactivation ratio, which was low with the cheap, powerful solenoid that we used. Initially, we relied on the activation/deactivation command in the G-code file, but users reported burning the solenoid, usually by leaving the software without safeguards. To address this, we made another change in the Marlin firmware. The M3 S1 command still activates the solenoid but only for 50 ms. This is long enough to emboss a dot and ensures that the solenoid is never activated longer than 50 ms.

We now have a G-code-programmable device to emboss Braille dots anywhere on a sheet of paper. To emboss a document in Braille, you can use AccessBrailleRAP. This software translates text into Braille characters, calculates the position of Braille dots on the paper, and then sends the

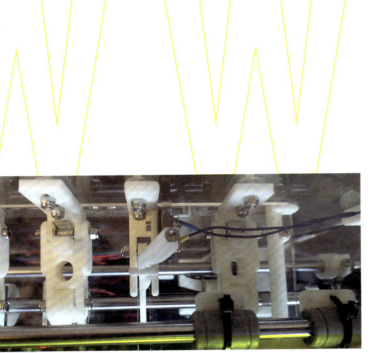

Left ←
Paper can be one of
the hardest materials
to work with because
of its tendency to jam
and crease

MORE INFO ABOUT BRAILLERAP

BrailleRAP website: **hsmag.cc/BrailleRAP**
Assembling guide: **hsmag.cc/BrailleRAPdoc**

G-code sequence to the embosser. However, you still need to translate the text into Braille. As we discussed earlier, there are different standards for Braille across the world, including variations between the UK and the US. As French lecturers, we started with basic software to translate text into Braille. However, we realised that there were many complex nuances involved in the process. For instance, marking a number is not the same as marking a group of numbers, and marking a capital letter is not the same as marking an entire word in capital letters. These details require careful attention to ensure an accurate translation. While searching online for an open-source Braille transcription library, we found Liblouis. It is a grammar interpreter that enables you to produce Braille translations using text files, removing the need to write code for a Braille translator. As Liblouis is open-source and

> ## "THERE ARE DIFFERENT STANDARDS FOR BRAILLE ACROSS THE WORLD"

publicly available, many contributors have provided data files for their own languages. Currently, Liblouis can translate Braille into more than 200 Braille standards. All the major languages of the world are supported, including and some regional ones. So, we stopped trying to code Braille transcription, used Liblouis, and focussed on providing accessible software to use BrailleRAP. □

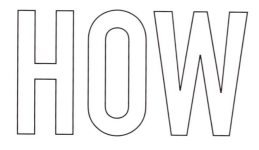

HOW + MADE

By **TIM RITSON**

DUAL AXIS SOLAR TRACKER

It all started with a free solar panel. I came across it in my late father-in-law's wardrobe, so I added it to the pile of things-I-might-never-use in the boot of my car. I knew he would approve of any way I could put it to use. Little did I know that it would be the start of a project that would become a mildly unhealthy obsession, starting small, but evolving into something… well … big and heavy! But awesome!

After bringing the panel home – an outdated, 20 W, 2.6 kg slab of silicon and aluminium – I started to think through what I might do with it. I decided to buy a battery and charge controller, and put it in the sun… just to see how it would go. I propped it up on an outdoor chair, but pretty quickly I had to keep on moving it out of the shade and rotating it to face the sun. I had recently been looking at

Left ←
Capturing the first rays of morning sunshine

DIY telescopes with Dobsonian mounts – a simple device not too different to a machine gun turret. It had two axes of rotation – the barrel of the telescope, like the machine gun, could tilt up and down, while the whole mechanism could rotate horizontally through a full circle. How hard could it be to build this for my solar panel, slap a couple of stepper motors on each axis to control the rotation, and run them from an Arduino?

And so it began. There was no real purpose to save the world with super-efficient solar generation. It was just simple curiosity. To see if I could do it, and where the journey would take me. I also figured it would be a good stepping stone to that DIY telescope – as is often said, why spend a few minutes aligning it manually, when you can spend several months automating it!

POWER SYSTEM

The first components I needed were to harness the power from the solar panel. My initial thought was to find an old car battery and hook it up. But, from some quick internet research, I learnt that a) old car batteries aren't that easy to find, b) aren't that cheap, c) aren't even the right type of battery, and d) you can't just connect a solar panel directly to a battery. It turns out, 12 V lead-acid batteries don't take kindly to being charged at 18 V.

Instead, a charge controller is the way to go, transforming the solar panel voltage to a suitable level for battery charging. I performed a few back-of-the-envelope calculations, and bought some appropriately sized components within my budget. And, with a little foresight, I chose a charge controller that had a USB output to provide an easy way to power the Arduino.

CONTROL MECHANISM

The next components needed were the motors to rotate the solar panel around each axis. I selected NEMA17 stepper motors, and 80:1 reduction gear-boxes. I had had some experience with stepper motors from a previous project (albeit much smaller and cheaper) and I liked their accuracy and 'digital' nature – they are designed to move through a discrete number of steps, so are well-suited to position control. The motors use TB6600 drivers, which receive 5 V, enable direction, and step signals and move the motor one step at a time.

Above ↑
First motor mounted

Right →
Familiarising myself
with the new motors

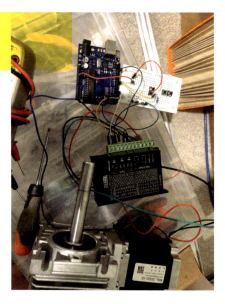

The overall mechanism is significantly overspec'd. The NEMA17 motor has 200 steps to rotate through 360 degrees – 1.8 degrees per step. However, upon receiving the motors and drivers, I hooked them up and began tinkering, and I soon discovered

the drivers also allow microstepping – dividing each step into 32 steps. So the motor can actually be controlled with 6400 steps – that's 0.05625 degrees per step. Now, combine this with the 80:1 reduction gear-box, and we have a cool 512,000 steps per rotation. I figured, speed isn't an issue, so why not? The more accurate the better! In astronomical terms, a resolution of 512,000 is equal to 2.5 arc seconds – less than the angular diameter of Mars (and pretty much any planet in our solar system)! If that doesn't mean much to you, imagine cutting a cake into 512 pieces, and then cutting each of those into 1000 pieces – a very sticky mess.

Anyway, as I would discover, other factors would constrain accuracy to be well outside of this range. And, in hindsight, it was also inefficient to use such a large integer – to encode 512,000, as you need at least 19 bits. So it required a 32-bit integer, where 16-bit would have been quite acceptable.

FRAME

In parallel with purchasing the motors etc., I was also logging into SketchUp online to create 3D models of various frame designs. This was helpful to visualise how all the parts might go together. I was limited to what I could purchase at a local hardware store – square tube and angles – and build with a relatively simple toolset – saws, drills, screwdrivers. I have tried to use aluminium as much as possible, partly to keep weight down and also because it doesn't corrode. I did experiment briefly with aluminium welding to fix the members

Below ↓
First section complete

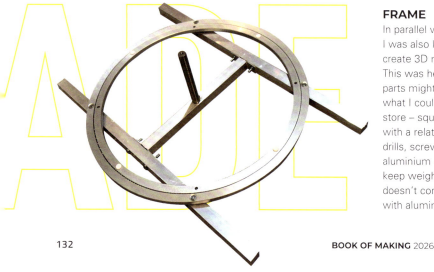

Below ↓
Fully assembled

together. However, it turns out that this is hard, and using a blow-torch to smelt aluminium on my wooden workbench in my wooden framed house, with wife and kids sleeping soundly above, just seemed like a bad idea. So, nuts and bolts it was.

One central part of the system was the rotating platform. It needed to support a significant amount of weight, while also rotating freely through 360 degrees. Initially, my research took me to some very nicely machined ball bearing housings, suitable for aerospace applications. But, having not won the lottery that week, I continued searching and soon came upon Lazy Susans – a fixture of any Chinese restaurant. Placed in the middle of a round table, they are spun around to allow the patron to serve themselves from the selection of dishes on the central turntable. A much better fit for my budget, and available from AliExpress in a wide array of sizes, I chose a sturdy-looking 16-inch version.

Piece by piece, section by section, it came together. I finally added the rotating tube to which the panel is fastened and added a counterweight to reduce the load on the shaft and motor. I needed a couple of specially machined parts for this from AliExpress, mostly designed for CNC machines. It wasn't pretty, but it was functional.

CONTROL SYSTEM

The brain of the device is an ATmega328P. Rather than use an Arduino, I wanted something more custom-built, so I designed and soldered my own circuit on prototyping board. The vital organs, if you will, are the push-buttons for manual control of the motors, a 16 × 2 LCD screen, and a Real Time Clock (DS3231). There are a number of headers for interfacing with the motor drivers, programming (USBasp) interface, USB power, ADC inputs, and additional I2C devices. Together with the charge controller and motor drivers, all this is housed in a weather-tight box – all cables enter through the bottom to avoid water ingress.

WORKING PRINCIPLE

Fundamentally, the operation is simple: the microcontroller calculates the elevation and

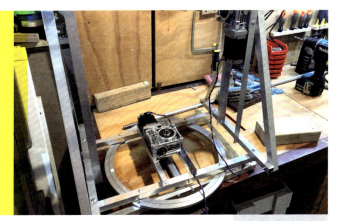

azimuth of the sun, based on its location on Earth and the date/time, and instructs each motor to rotate until the panel is at the desired position.

First, the program also needs to get the initial position of the motors in order to know how far to rotate. Unlike servos, stepper motors are 'open loop', meaning they do not provide any position or speed feedback, so it's necessary to keep track of the motors' positions manually. This was initially achieved by manually rotating the panel (via the push-buttons) to a known position the first time the device was switched on. I then programmed it to save this position in EEPROM so that it would remember its position when power is switched off.

This would require EEPROM to be written any time the motors moved. For thoroughness, I even implemented EEPROM wear levelling. This is a method to avoid surpassing the rated number of write cycles of EEPROM (100,000) by writing data (the saved motor positions) to a different memory address each time.

Next, date and time is received from the RTC. Latitude, longitude, and time zone are hard-coded. Sun position calculations are based on David Brooks' open-source code from 'Arduino Uno and Solar Position Calculations (dated Feb 2015), **hsmag.cc/ SolarCalcs**. I'd like to say I developed my own calcs, however, I believe in standing on the shoulders of those who came before us (and also because I didn't have time to get a degree in astronomy).

A timer interrupt was used so that the program would periodically wake, calculate the sun position, and then rotate the panel. I typically set this to happen every 60 →

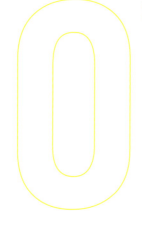

Above ↑
Vertical sections installed

Below ↓
It was a very tight fit getting all the electronics in one space, while also keeping connections accessible

Dual axis solar tracker

Right →
The three-axis magnetometer, located as far as possible from sources of magnetic interference

Below ↓
The uncalibrated magnetometer readings. The azimuth is calculated via arctan (x,y), effectively drawing a vector from (0,0) to a given point

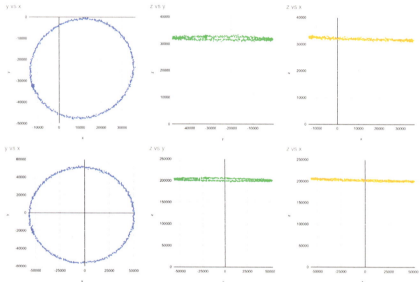

Above ↑
After calibration, the blue circle is centred at (0,0)

> ## "THE PROJECT HAS BEEN A CONSTANT LEARNING EXPERIENCE"

calibration process unnecessary. I installed a magnetometer (QMC5883L) to measure azimuth (effectively a compass heading) and an accelerometer (MMA8452Q) to measure tilt. When first received, the magnetometer didn't work well. I realised it needed calibration. The Earth's magnetic field is weak, so the magnetometer is sensitive by necessity. However, this means it is susceptible to external magnetic fields. With a large lead-acid battery and two DC motors nearby, it wasn't exactly an ideal environment for it. So, I placed it on top of the frame, as far away from these as possible, and calibrated it in situ.

Edward Mallon's blog post was a godsend for figuring this out: **hsmag.cc/ EMallonCavePearl**. This explained the process to plot X and Y readings from the magnetometer (and Z too if needed), which trace out a circle as the magnetometer is rotated. The circle's centre should be at 0,0, and the circle should be round. They were way off initially, so I used a free piece of software (as explained in the blog) to calculate a set of corrections that are hard-coded into the azimuth calculation.

CHALLENGES
The project has been a constant learning experience, discovering issues and figuring out solutions as I go. The first problem was a reminder that aluminium is soft. The shaft from the azimuth motor is fixed into a square aluminium tube in the base. The shaft has a key, and I cut a corresponding notch into the tube. However, after a while of operation, it fully reamed out the hole.

seconds. However, this can be set faster. Another method would have been to constantly rotate the panel at a controlled speed. However, this would require more complex calculations to time each motor step – the speed of each motor would be different and vary constantly throughout the day.

Each time the program wakes, it must also check if the sun is actually up, i.e. if the time is between sunrise and sunset. Sunrise and sunset times are calculated as part of the sun position calcs, as they differ throughout the year and by location.

When night-time is reached, the code changes to 'night mode', whereby it rotates the panel to the starting position for the next day, and waits, patiently, until the sun rises again.

SENSORS
Most recently, I have added two sensors so that the system can measure its own position. This makes the manual

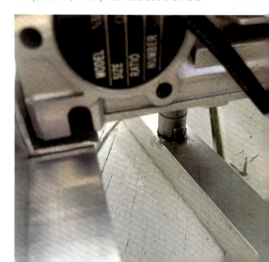

HOW

Luckily, I had a spare part made of steel that fitted over the tube snugly and had a properly machined key. I also found that the movement of the panel was jerky. The heavy panel is fastened directly to the rotating tube, so there is almost no leverage, causing it to wobble easily. To mitigate this effect, I programmed the motors to ramp speed up and down gently when starting and stopping, smoothing out the jerkiness.

In hindsight, I would build this differently next time. Perhaps using a freely rotating shaft, controlling the pitch via an arm attached to one end of the panel. I also noticed that there is quite a bit of play in the gear-boxes, a bit like a loose steering

Above ↗
The part I used was a bit of a hack – half of a shaft coupler that just happened to snugly fit over the square tube

Left ←
The damaged hole in the soft aluminium square tube caused by the shaft key

wheel on a car. When the motor driving the gear-box is rotating in one direction, the output shaft rotates correctly, but when the direction of the motor is reversed, the output shaft is loose and doesn't engage immediately.

To overcome this, I measured how much play there is, and programmed the motor so that when it changes direction, it moves through an additional few degrees to engage in the reverse direction.

Unfortunately this is a case of 'you get what you pay for'. Gear-boxes with the accuracy I had hoped for would be a great deal more expensive. The program is also nearing the memory limit of the microcontroller. I first noticed this when I

wrote a function to simulate one full day of movement. It calculated position data for every few minutes of a day and saved it into an array. I had to reduce the size of the array or the program would crash. I also discovered that text strings use a lot of memory.

So, all the serial debug messages and LCD output strings were using up the majority of available program space. I eventually implemented some preprocessor code (e.g. `#ifdef DEBUG_ENABLED #endif`) so that debug messages wouldn't compile unless explicitly enabled in code.

Above ↑
The 'finished' product

CONCLUSION

No project is ever 100% complete, and this project is no different. I still plan to measure the power generated by the solar panel, and possibly the power drawn by the tracking mechanism. By comparing the two and checking against a non-tracking scenario, I will find out if this has actually achieved anything at all – that is, does it generate more power than it uses by rotating to track the sun? It doesn't really matter either way – I now know so much more than when I started. I've learned a lot of technical skills, learnt some lessons, but also that I can achieve a lot by breaking a problem down into manageable chunks and working through it. I will certainly need this patience for the next stage – building a telescope! □

HOW I MADE

By **DR. ROSS HAMILTON** with **ROB MILES**

MIDI PLAYER PIANO

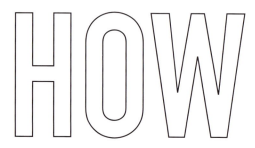

t turns out that you don't actually decide to create a digitally controlled piano containing 88 solenoids, a bunch of controllers, a beefy power supply, and an awful lot of wires. You just kind of end up with one. I really like the sound of a live, well played, piano. So, I thought, why not learn to play one? I was lucky enough to have a piano at home, and so I loaded up a piano teaching app and got started. The app listened to the 'music' that I was making, and told me if I was playing it right and making progress as a pianist.

And then the problems began. It turned out that piano teaching apps are not good at recognising what they hear once the music gets complicated. Pieces I played perfectly apparently contained bum notes, and I could easily defeat the app and get a perfect score by bashing lots of keys at once. This is not a criticism of the app: recognising individual notes in a bunch of simultaneously played ones is fiendishly difficult, and doing it consistently is hard too. And it does, mostly, work. However, an accuracy of 99% still means that one note in 100 was not recognised correctly, making practice more work than it should be. So, the next step was to find an electronic piano with a MIDI (Musical Instrument Digital Interface) connection that could send note information directly to my tablet. So, I started searching for a suitable instrument.

Figure 1 ←
The brand name probably reflects the fact that many pianolas were used in churches to provide music for hymn singing

Figure 2 ↓
Some rolls are also printed with music lyrics and sound effect instructions

Figure 3 ←
The coins were added to the keys to give them weight and make them move down when their note is being played

PLAYER PIANO

It turns out that when you search for 'digital pianos' in an online marketplace, you also get bombarded with listings for other things, including mechanical player pianos or 'pianolas'. These are awesome machines I'd not seen much of before. They are like a piano, but heavier (as if you thought that was possible) because they contain a lot of extra machinery. They have been around since the early 1900s, and were one of the first devices that could play recorded music (the other one being the phonograph).

You load your pianola with paper rolls which have tunes encoded on them in the form of little holes, and a mechanism inside the piano plays the tunes for you. Very interesting. And, when you find that there is a pianola available for free just down the road, what do you do? Answer – you borrow a trailer, drag the pianola home (plus a goodly number of music rolls), and then become enthralled with the sound that it makes. So, part Western saloon, part Victorian parlour.

Figure 1 shows my first pianola. The hole in the middle is where you load the music roll. Then you pump the pedals and the piano plays itself. You can control the speed of the playback and the volume of the sound using the controls on the front of the piano. You can use it as a normal piano too. **Figure 2** shows a roll of

piano music: '*Hawaiian Nightingale*'. The legend 'Hand Played' on the label means that the music was recorded from a live performance rather than transcribed onto the paper from the written musical score. Hand-played music sounds better because a human player can add more expression and 'syncopation', which means playing the notes slightly off the beat.

Figure 3 shows the insides of another pianola I seem to have acquired (they are addictive). This one needed some serious pneumatic repairs, as the brightly coloured pipes will testify. The three vertical panels on the right-hand side of the piano are driven by a crank at the top, and connected to the bellows which power the note playback.

The music roll is powered via a complicated set of gears and chain drives, and pulls the paper strip over a reader in the middle of the piano. The reader contains 88 holes, one for each note. Normally, each hole is blocked by the paper in the music roll. When a hole in the paper goes over the reader, however, it lets air into the reader, triggering a tiny set of bellows which moves the piano mechanism for the note controlled by that track. →

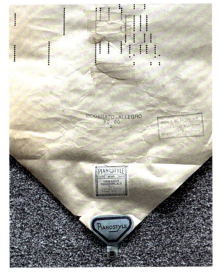

FEATURE

Figure 4 ↙
This might look simple, but the video of it did attract several thousand views

THE PIANOLA LIFE

So, now I have multiple pianos, including one which will play itself. This is fun! Although, a pianola is a high-maintenance device. I must do all the piano-related fixes, making sure that the strings are tuned to the right notes, that the felts (which damp the strings when a key is released) are in good condition, and the keyboard mechanism works correctly. But I also need to deal with the huge number of tiny rubber tubes and bellows which transmit the air pressure to trigger the notes, and then ensure that the drive mechanism, which pulls the paper roll through the pianola, is working correctly and moving the paper at 70 inches per minute.

Eventually, after a lot of work, the pianola becomes a part of the furniture, and I enjoy working my way through the music rolls that came with it, hearing songs that might not have been played for many years. And then I start to wonder if there is a way of making a piano which could play MIDI files directly, rather than needing specially made paper rolls (although a surprising amount of contemporary music is still released on this format). So, why not convert a piano to understand MIDI notes? And so, that's what I did.

MAKING A MIDI PLAYER PIANO

Figure 4 shows my first experiment with a computer-controlled keyboard. This involved using servos to play piano notes. This was discovered to be workable, in that you can use the arm on the servo output to make a piano key move, but the movement is too slow to be properly usable. So, the next step was to look for some suitable solenoids.

A solenoid is made up of a coil of wire and an internal core which is pulled into the coil when you pass current through the latter. I use this movement to 'press' a key, although there are some slight complications. The solenoids I'm using have coils with an internal resistance of 8 ohms and are powered by a 12-volt supply. If I do the sums with Ohm's law, I discover that each coil draws 1.5 amps. A standard piano

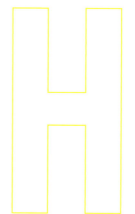

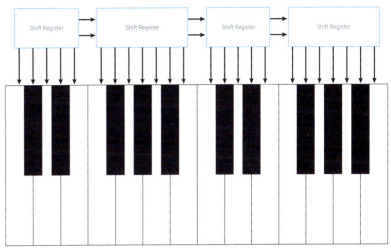

Figure 6 ↑
Three shift registers can control two octaves

has 88 keys, which means that in a worst-case scenario, with all the notes played at once, the piano would draw more than 120 amps at 12 volts. I decide not to think about this too hard and order a 50-amp, 12-volt power supply on the basis that no more than 30 or so notes will be sounding at the same time. After all, a piano player only has ten fingers…

CONTROL THE POWER

Now that I have a power supply and some solenoids, I need to think about how I can control the power for each note. The good news is that there are circuits online to help with this, and the better news is that colleagues at the office are keen to help with the creation of circuit boards that can be used to control a bunch of notes. We design a circuit board and get some made, and we also order all the components.

Figure 5 shows a completed driver board which can drive 16 solenoids. There are six of these boards in the piano, connected in a 'daisy chain', with the output of one board connected to the input of the next. The components at the bottom of the board in **Figure 5** are 8-bit shift registers which hold the signal states to be sent to the MOSFET drivers which control the solenoids.

SHIFTY BUSINESS

Figure 6 shows how the shift registers are loaded with note information. Each shift register has 'clock' and 'data' inputs. The microcontroller sends the note data into a register by moving a clock signal up and down. When the clock signal rises from low to high, the shift register samples the input data and stores it in one bit. When the clock signal falls, the register shifts its contents along one bit. The result of the waveform in **Figure 6** would be to set the first four keys to the pattern '1010'. In other words, the keys C and D are pressed. A shift register also has a clock and data outputs which can be passed on to another shift register to allow longer bit patterns to be stored. After 88 clock pulses, all the registers would be loaded with a pattern of signals that represents all the keys on the keyboard.

Each shift register also accepts an 'output enable' signal. When this signal is set low, the register will output its data values onto the output pins of the chip. These signals are used to control MOSFET switches which send power to the solenoids. There is also a 'reset' signal which can be used to clear the contents of the shift register.

Each time the keyboard needs to be updated, the microcontroller sends a new set of clock and data signals, and the new values are then latched onto the outputs.

QUICK TIP

It turns out you can pick up second-hand pianos (and even pianolas) for tiny amounts of money. In fact, people are so keen to get rid of them that if you offer to go round and pick it up, you can often get it for free.

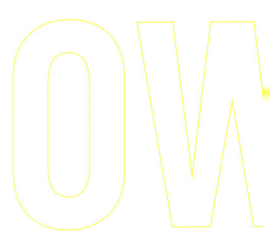

FEATURE

Figure 7 ←
The last half of the final driver board is not used as there are only 88 keys

Figure 8 ↓
The hammers for each note are above the shiny metal bar going across the middle of the picture

MAKE A NOTE OF THIS

Figure 7 shows the complete player piano electronics. The solenoids are mounted on aluminium carriers. Each solenoid has an actuator which goes up through a hole drilled in the piano. When the solenoid is triggered, the actuator moves up through the hole and pushes against the underside of the piano key.

Figure 8 shows how a piano works. Each piano key is a tiny see-saw. When you press down on the front of a key, the back of the key goes up, hitting a mechanism causing a hammer to hit a string and make a note.

When you let go of the note, the back of the key goes down, releasing a small felt pad which then rests against the string, stopping the sound. It is quite a complicated mechanism, and it is repeated for each of the 88 notes on the piano keyboard. It is said that a piano contains 12,000 different parts. Looking at **Figure 8**, you can begin to see how this is the case. The electronic piano uses the actuator from a solenoid to push the back of the key up and sound the note.

Figure 9 shows the piano keyboard with four keys and the hammer mechanism removed. The figures shows the actuators on the ends of the solenoids. When the actuator goes up, it hits the bottom of the key and moves it up, playing the note. The holes had to be drilled in a staggered formation because the width of each solenoid was greater than the width of a piano key.

MIDI OF THE ROAD

The final link in the chain is the microcontroller, which provides the MIDI interface to the outside world and generates the data signals for the shift registers. **Figure 10** shows the controller board, along with one of the solenoid controller boards. The clock and data signals for the shift registers are generated by an Arduino Pro Micro. This device supports 'USB hosting', which means it can appear to an external device as a MIDI peripheral and accept note on and off messages. The ESP32 device above it on the circuit board is used so that the piano can also be used via Bluetooth MIDI.

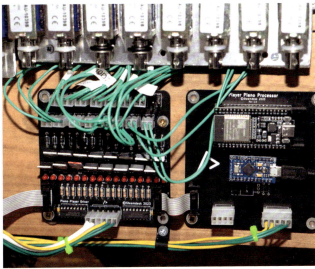

Figure 9 ↑
Each key needs to be individually adjusted so that the keyboard doesn't look 'snaggle-toothed'

Figure 10 ↗
The controller board is on the right of the picture

Figure 11 ↘
The piano is always a hit when I take it out to demonstrations

THE THING LIVES!

The piano is becoming a bit of a star with quite a following on YouTube. **Figure 11** shows it being demonstrated at the Rural Lincolnshire Enterprise Hub. It has acquired some remote-control facilities, courtesy of an Amazon Alexa interface which lets you ask for tunes by name and have them played. I've also been looking at using generative AI to create music with a particular style on demand. It would also be interesting to find a way of reading pianola rolls so that they could be converted into MIDI notes to be played on the piano. And finally, I'm exploring adding coloured lights to the keyboard. Everybody loves coloured lights.

If you want to follow along with my journey, you can find the pages for this project at **playerpianos.co.uk**. There are videos of the piano in action, more of the story behind it, hardware designs for the interface printed circuit boards, and even a place to chat about piano stuff.

The project has been a lovely example of how you can start with something simple (I'd really like to learn a bit about the piano) and end up with a working knowledge of piano history, get a mechanical pianola (or two) of your very own, and then branch out into electronic control, the MIDI protocol, and even artificial intelligence. It's been great fun, and I think we have only just started… ▫

ULTIMATE WORKSHOP

By **ANDREW LEWIS**

Every maker dreams of having their own workshop. Even those fortunate enough to already have a space of their own wish that they could change a few things to make it perfect. It turns out that a perfect workshop isn't a static thing. It's a living space that changes as your skills advance, and as the focus of your projects evolve. Even though a leather-working studio is going to look a lot different to a metalworking space or an electronics lab, there are areas of overlap between the tools and methodologies employed by almost every maker, and that's what this article is about.

There are two ways to get the perfect workshop – you can do it by accident or by design. In the former case, you start with a largely empty space and, as you complete projects, the space fills with the tools and materials that are most useful to you.

You'd typically see this type of workspace in an old, established firm producing handmade goods. It's the sort of workspace where you don't touch anything, and you can't find anything unless it's your space. It can be a comfortable way to work, but it's almost impossible to move to a different space if you need to, and it's impossible to work with someone else because you'll have to keep finding the tools for them because only you know where everything is. Chances are, safe working practices in a space like this involve squinting when you're welding, and pouring a cold cup of tea over any fires that break out. Alternatively, you can take a more structured approach to your workshop design, take a look at other people's workshops, think about how you want to work, decide on the tools that you need, and then design a workshop accordingly. This is by far the most sensible approach to creating a workshop.

BUILD A MODEL

Assuming you have acquired a suitably sized space to work in, the most important thing you should do is make a 3D model. It doesn't matter if the model is made from cardboard, 3D-printed plastic, or is modelled virtually. Getting an accurate feel for the empty space will let you make good decisions about layout in a way that a 2D map can't.

The most valuable commodity in a workshop is space. It's very easy to fill your workshop with tools and equipment to the extent that you don't actually have any space to work in – so think of the largest item you're likely to want to make, and make sure that you leave enough space to move around that object with any equipment you need. Consider dividing your space into virtual zones or work areas so that you know what sort of task will happen in a particular place. That way, you'll be able to make better decisions about the routing of cables, ventilation, and storage for that zone. Your 3D model can be very helpful for this.

Do not keep infrequently used specialist tools or large amounts of scrap materials in your workshop. Either rent tools as you need them, or store them off-site. If you have leftover materials from a project, dispose of them unless you are absolutely sure you will need them in the near future. It's helpful to remember that empty space costs money, and disposal doesn't mean destruction. It might be possible to sell or recycle any leftovers from a project, or even store them off-site if you have space somewhere else.

QUICK TIP

Put wheels on everything. Your needs might change in the future, so put wheels on everything that you can. Being able to move a surface or storage wall easily can be very useful.

Above ◆
Splitting your workshop into clean and dirty zones can help prevent contamination of sensitive equipment with dirt or oil. Some tasks, like grinding or welding, should never be done in a workshop with a clean area – they should only be done outdoors or inside a dedicated workshop

CONTROL THE POWER

Most people with a home workshop have a list in their head of what pieces of equipment they can use together without tripping the breaker or blowing a fuse. It's not unusual for a shed or outbuilding to have a single 16 A circuit for the electrical sockets and lights, and that isn't a lot to work with when you consider things like electric heaters or kettles, laser cutters, compressors, and welders. Ideally, you'll have an electrician connect 30- or 40-amp circuits to your workshop with a choice of single- and three-phase power, and you won't have to worry about what tools you run. However, unlimited control over your power delivery isn't always possible, and there are a couple of things that you can do if you're struggling.

Fit larger tools with soft-start circuits if possible. For a lot of large machines like air compressors, the biggest power draw occurs when the machine first starts, as motors overcome inertia and friction in the system. Once the machine is running, the power draw is less intense. A soft-start module slowly ramps up the power delivery to the machine, avoiding the initial spike that might trip your power breaker. You can separate out non-essential electrical devices like electric heaters and wire them all onto a circuit with an isolator switch. That way, you can temporarily switch off all of the non-essential devices at once, maximising the amount of available power for other machines in the workshop. Critical machines like laser printers and computers can be given their own UPS (uninterruptible power supply) so that jobs won't fail if the power breaker does trip.

In the case of something like an air compressor, a larger tank will mean that the motor needs to run less often. A large tank makes it possible to complete a job with the compressor motor turned off, then refill the tank at a later date. This can also reduce noise levels if you're forced to work late. →

QUICK TIP

Always have emergency lights in a workshop. You don't want to be stuck in the dark with potentially dangerous equipment all around you.

Above ◆
In the UK, messing with your consumer box or electrical systems isn't acceptable and will probably invalidate your household insurance unless you're a qualified professional. If you have to run wires to your workshop, get a professional to check the work and connect it to the consumer unit

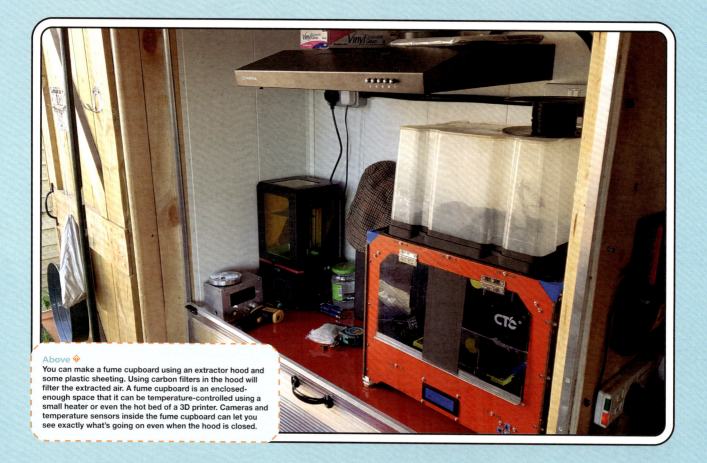

Above ◈
You can make a fume cupboard using an extractor hood and some plastic sheeting. Using carbon filters in the hood will filter the extracted air. A fume cupboard is an enclosed-enough space that it can be temperature-controlled using a small heater or even the hot bed of a 3D printer. Cameras and temperature sensors inside the fume cupboard can let you see exactly what's going on even when the hood is closed.

LET THE ROOM **BREATHE**

A workshop should be a comfortable space for you to work in, but it should also be comfortable for your equipment and consumables. Air quality, temperature, and humidity are all important and need to be controlled in a workshop. The amount of volatile organic chemicals (VOCs), dust, and other particulates must be kept to a minimum to reduce the chance of long-term health problems like uncontrolled explosive defenestration, nausea, headache, and death.

Use a dedicated fume cupboard with active extraction for hazardous jobs, and use extractor fans with ducts to remove dust, smoke, and fumes while you work. Be careful about venting your fumes into the open air. Fit activated carbon filters in your system and change them regularly to reduce airborne pollutants. Check that the extraction system is working effectively with an anemometer, and regularly check the VOC levels inside and outside your workshop.

A shop-vac is a good way to reduce airborne particulates, and many power tools have adapters that allow you to fit a vacuum hose to them. You can reduce the need to empty the shop-vac by adding a cyclonic filter or dust separator. The cyclonic filter sits between the vacuum cleaner and the machine you're using, and separates larger particulates into a sealed container. While they're more suited to built-in vacuum systems, it is possible to add them to portable vacuum cleaners. To reduce particulates even further, a dedicated filtration system like the Scheppach 100W (**hsmag.cc/Scheppach**) can filter the air. If you're working regularly with MDF or powdered materials, a filter system like this is essential.

A workshop should always be kept near room temperature. Letting some equipment/consumables get down near freezing temperatures can spoil them. Pipes can crack, oil will thicken, and paint will separate. Laser tubes can shatter if they are started too cold, and metal rails can buckle. A radiator with a frost setting should be enough to keep most insulated spaces out of the danger zone. An active temperature controller can keep the coolant in a laser at a preset temperature, ensuring that the tube won't freeze.

QUICK TIP

Lighting is one environmental factor that gets ignored. You need more light. Good overhead lighting, desk lamps, and work lights make a huge difference.

LET'S PLAY:
KEEP, STORE, DISPOSE

There's an old saying about storage in the workshop: "A workshop isn't a storeroom, and a workbench isn't a shelf." While it's OK to store some commonly used tools and consumables in your workshop, don't make the mistake of letting the space become something that it isn't supposed to be. The more things you store in the workshop, the less space you'll have to work. Adam Savage talks about storage in his Tested series, specifically the idea of "first-order retrievability" – where the tools and materials you need to complete a task are visible and easily retrieved. To paraphrase Adam further: "drawers are where tools go to die."

This is true, but can also be thought of as "If you can't find it, you don't have it." In a smaller workshop, it's necessary to hide things from view to maximise your available space. It isn't a big problem if you indicate clearly where things are stored using labels.

Storing tools vertically is a great strategy in a workshop, provided it doesn't cause a safety problem. Having tools on hooks above chemical tanks or machine tools is an industrial accident waiting to happen.

There is no one-size-fits-all solution for materials storage, but the Stanley Fatmax Waterproof Pro Organisers (**hsmag.cc/SF_Organiser**) are great for smaller components, and the 'deep' versions of the organiser can hold larger items like stepper motors and batteries. The Recycled versions of the Really Useful Box (**hsmag.cc/UsefulBox**) are great for large items, and can be labelled easily with a DYMO labeller. For portable storage, the Hobbycraft three-tier storage trolley is excellent (**hsmag.cc/StorageTrolley**) and is low enough to slide under most desks when it's not in use. →

QUICK TIP

If you don't want to use a tool because it's too difficult to get to, then you might as well get rid of it.

PLAY IT **SAFE**

The most important part of every workshop is safety. Even if the workshop is designed entirely for one person to use, you still need to consider the safety of others when you are working. The use of goggles, gloves, earplugs, and machine guards is well described elsewhere and doesn't need to be repeated here. Hopefully most readers will understand that working alone is dangerous, and remember to tell someone when you should be finished working. There are other considerations that you should make to improve the safety of your workshop. Adding a door alarm or a bell to your workshop door will let you know if someone comes in unexpectedly, and adding a warning light outside the door will warn people when you're doing something dangerous and probably shouldn't be disturbed. Adding a lock to the door is a risky proposition if you are the only person inside the workshop that can unlock the door. If you are injured, help might not be able to get to you through a locked door.

You should always have a well-stocked first aid kit with an eye-wash kit, and there should be enough fire extinguishers, fire blankets, sand, absorbent pads or sawdust to handle fires and chemical spills. Important safety equipment needs to be easily accessible from the place that it will be used, and you should have spare items to cover breakages and possible visitors.

Something often overlooked in the workshop is the idea of keyed access or interlocks to larger machines. If you add an electronic key to a machine, you can effectively control who can start and use that machine, and you can also use physical locks to mechanically isolate the start buttons of a machine. Adding a key to a machine isn't usually difficult, and the safety benefits you will get from the addition are enormous.

Danger signs are an important defence against injury through ignorance. If you have dangerous tools or materials, make sure that you use the appropriate signs to warn people of potential dangers. Remember that signs and other safety features aren't just there for your safety. They are there for the safety of anyone who enters the workshop.

TOOLS **GLORIOUS** TOOLS

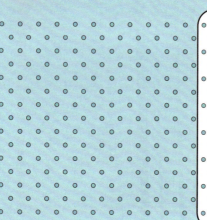

It's almost impossible to recommend the perfect toolkit, because different interests require different tools, and there are many ways to get the same result with different tools. However, there are quite a few tools that should be considered as essentials for everyone's workshop and a few that are considered as 'nice to have.'

As a general rule, having two sets of every important tool is good for several reasons. If you break a common tool, you've got a spare of that tool to work with and you can carry on. You can also keep a 'dirty set' of tools that you use for messy jobs, while the other set remains clean.

YOU'VE **GOT THE POWER**

A basic set of power tools in combination with the right sort of jigs can duplicate most of the capabilities of a workshop full of large-scale machine tools. Note that the key word here is capability, and not capacity. If you are running a factory making wooden benches, it makes sense to have a floor-mounted thicknesser and jointer because they can process industrial quantities of material very quickly. For general making, simple power tools are usually enough.

It's good to get tools that share a common branding so that batteries can be interchanged without needing an adapter and multiple chargers. Every maker will have their own favourite range of tools, with Triton, Milwaukee, Bosch, and DeWalt being popular names in the field. DeWalt kits usually represent good value, and spare parts are easy to get. With batteries that slide onto the bottom of the tool's grip, many DeWalt tools have slightly smaller and better-contoured handles than their competitors. This makes them a more comfortable choice for makers with smaller hands. For the average workshop, a drill with a reverse and hammer function, a jig-saw, and a multi-tool will be the essential kit that you'll use to do the majority of your making. If you do more metalworking, an angle grinder and portable bandsaw will be useful. If you do more woodworking, a router and circular saw will probably be more useful. It's worth noting that tools like routers and circular saws can often be used to cut aluminium with the appropriate blades. Cutting steel usually requires a separate saw with a lower-speed motor.

Belt and orbital sanders are good for shaping and finishing materials quickly, and can often be bench-mounted, giving them a similar functionality to a belt sander. Additional power tools will give you very specific advantages in certain situations but probably aren't necessary for everyday working. An impact driver is a common power tool these days, and it's excellent for driving in screws. However, a drill can do the same job and, in certain materials like concrete, the impact driver would shatter the material. Right-angle drills are very useful if you're working in a tight space, but are unwieldy in normal use. The same applies to other tools like die grinders and mouse-style electric sanders. →

QUICK TIP

A hot-glue gun is a power tool, and it deserves more love than it receives. Get a hot-glue gun, and thank yourself for years to come.

LEFTY LOOSEY, **RIGHTY TIGHTY**

When it comes to screwdrivers, there are a mind-boggling number of choices to make. For electrical screwdrivers, it's difficult to fault Wera's offerings. Its Kraftform fixed (**hsmag.cc/KraftformFixed**) and interchangeable (**hsmag.cc/KraftformChange**) screwdrivers are precise, strong, and ideal for general workshop use.

> For smaller items, Wera offers a basic Kraftform Micro Precision set (**hsmag.cc/KraftformMicro**), although for the serious maker who's looking to repair modern electronic devices, it's difficult to ignore the excellent iFixit Manta Precision set (**hsmag.cc/iFixitBitSet**).

> Spanners are another tool where there are literally thousands of options, but for general open spanners, the Draper Expert series (**hsmag.cc/DraperExpert**) offers a reliable spanner with a handy coloured finish that makes it easy to identify different sizes.

> Wera's Allen keys (**hsmag.cc/WeraAllen**), socket sets, and Torx wrenches have a similar colour-coding scheme that makes it very easy to pick the right tool when you have several on the desk.

Above
It's often said that you should buy the cheaper version of a tool first to see whether it suits your workflow, then get the higher-quality tool if you find it useful. This might be true for unfamiliar tools, but for ubiquitous items like screwdrivers and spanners, go for the expensive tool first. These are tools you'll always have a use for, and buying quality pays off

ELECTRONIC **DREAMS**

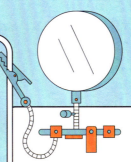

A lot of nonsense and snobbery exists in the electronics world about which tools you should be using – especially when it comes to multimeters. If you're working in an industry where a tenth of a volt is enough to ruin your plans for global domination, then you should really look into buying a very expensive set of calibrated tools. If you're an enthusiastic tinkerer or working on general repairs, a basic multimeter is going to be absolutely fine.

Thirty years ago, the first item to land on an electronics workbench after the soldering iron and multimeter would be a CRT oscilloscope. These days, an oscilloscope is much smaller, much less expensive, and works over a USB connection. For a basic unit, something like the Hantek USB oscilloscope (**hsmag.cc/HantekOscilloscope**) is fine for everyday use.

> If you're likely to be working on battery packs, a battery welder (**hsmag.cc/SeesiiWelder**) is absolutely necessary. It might appear to be a niche product, but it offers a solution that nothing else can really achieve. Trying to solder a LiPo cell with a conventional soldering iron will damage it, so mechanical contacts or welding are the only real choices you have for joining cells into a pack.

> For soldering, an SMD rework station that combines a heat gun and a soldering iron (**hsmag.cc/KatsuWelder**) is great for desktop use. The air gun is an extremely useful tool to have on the desktop, making fast work of small jobs like applying heat-shrink tubing or bending plastic pieces to shape.

> A USB microscope and borescope are useful additions to a soldering station, particularly if your eyes aren't as sharp as they used to be. Seeing clearly into confined spaces can make all the difference to a soldering job.

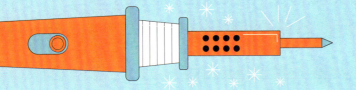

TOMORROW'S TECHNOLOGY TODAY

We are fortunate enough to have access to a few tools that simply didn't exist when our parents or grandparents would have been building their workshops. An obvious example of this is the 3D printer, which is one of the most versatile workshop tools if you know how to use it correctly. While desktop 3D printers typically produce plastic or resin parts, their strength isn't really in the production of finished parts. Their true strength lies in their ability to create jigs, frames, supports, temporary tooling, and moulds for other parts that radically simplify the processing of other materials. You can create moulds for casting, jigs and templates for accurate drilling and cutting, or even just shims and spacers to keep parts accurately positioned.

Laser cutters are similarly useful for creating short batches of repeated shapes, engraved designs, or accurate cuts on flat sheets. If you have a need to do this sort of activity on a regular basis, a laser cutter is a superb (although very expensive) addition to your workshop.

Lastly, the often overlooked workhorse of many workshops is the computer. Having reliable computers in your space to control machinery and do general work on is incredibly important. The best advice about computers is to keep multiple off-site backups of important data, and don't put all of your effort into building a single computer that does everything. Keep your computers separate so that a single failure doesn't take down the whole workshop. □

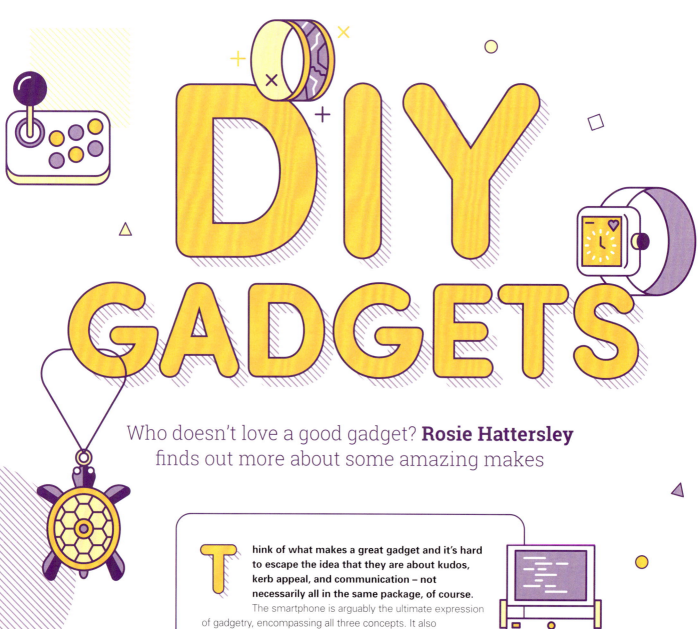

DIY GADGETS

Who doesn't love a good gadget? **Rosie Hattersley** finds out more about some amazing makes

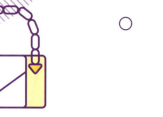

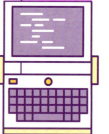

Think of what makes a great gadget and it's hard to escape the idea that they are about kudos, kerb appeal, and communication – not necessarily all in the same package, of course.

The smartphone is arguably the ultimate expression of gadgetry, encompassing all three concepts. It also demonstrates the importance of portability. How else to gain that all-important kudos without being able to take it out and about to show it off to our friends? We kick-start our look at the world of homemade handhelds and wearables with two great phone examples of home-brew design before applauding repurposed tech to create incredible cyberdecks and tiny PCs. Electronics play an important part in musical and visual gadgets too, as well as in homages to iconic films and games. With cosplay becoming ever-more indistinguishable from original props and costumes, we highlight some amazing makes along with achievable accessories you can 3D-print or simply customise.

PHONES

Keep in touch or upcycle some old tech

OURS smartphone

If you're taken by handset design rather than upcycling an existing phone, you could do a lot worse than follow Evan Robinson's build instructions for the 4G LTE OURS (open-source, upgradable, repairable smartphone). Running Raspberry Pi OS and primed with familiar smartphone apps such as Facebook, YouTube, and WhatsApp, it has a 5MP camera, microphone, and volume control, as well as an HDMI port to attach an external display alongside the OURS' 4-inch Waveshare touchscreen. Evan created the handset with privacy and security in mind, but it is also great value, with a bill of materials costing roughly 180 euros.

Evan says the most difficult part of the build was getting the power system stable enough to power the machine through boot because it uses a lot of current upfront, and sporadically for activating the 4G LTE module, but then uses less current once the OS is loaded. He spent a lot of time fiddling with tiny potentiometers on voltage regulators, then turning the system on and off over and over again. Since unveiling OURS, the idea has been embraced by dozens of other makers, with plenty of constructive feedback and around 40 enthusiasts actively working with Evan to develop a new model.

hsmag.cc/OurPhone

Antique telephone

Maker Mark Lister bought an antique phone from eBay, took out its existing electronics, and added a Raspberry Pi 3B from 2022 with bespoke GSM HAT and a speaker. The Raspberry Pi also provides internet access. Mark mapped the numbers on the old phone's rotary dial to Raspberry Pi's GPIO pins to create a usable handset that stores numbers and reads out the name of callers it recognises. The SIM card in the GSM HAT recognises the pulses generated by turning the dial as digits, thanks to Mark's specially written code. Further features include a backup UPS battery and, of course, a proper ring. Brrring, brrring!

hsmag.cc/OldPhone

Phone Map

If you're holding a phone handset up to your ear, it might as well deliver something you want to hear. Caroline Buttet decided she'd like to enjoy music from around the world, using the **radiooooo.com** online radio station player, an Arduino, and Raspberry Pi to serve up broadcasts, depending on where on a map of the world she placed her pin. The user can also pick up the rotary phone handset, dial the country code, and select a decade to choose what they hear. Caroline used Raspberry Pi to connect to the radio player app, while an Arduino hidden in a metal panel behind the framed world map controls the country selector.

hsmag.cc/PhoneMap

CLAMSHELL BLACKBERRY CYBERDECK

Because everything's better when it's tiny

The original business smartphone, BlackBerry's main legacy now seems to be the usefulness of its tiny keyboard in retro builds and cyberdecks such as this laser-cut, wood-clad version by Michael Klements. "Most keyboards, even compact and foldable ones, are three to four times bigger than the final build size that I was going for – and you'd still need to add a mouse to that," Michael explains. However, he was delighted to discover Solder Party's BB version on Tindie (**hsmag.cc/BBQ20Keyboard**). The Clamshell Blackberry Cyberdeck features a HyperPixel 4.0 display which hooks up to Raspberry Pi 4 via its GPIO pins. The advantage here is that there's no need for an additional power source for the screen, meaning Michael could keep his build ultra-compact. This aim also involved a tricky design process, so the hinged plywood case, sketched out in Inkscape, is just large enough to accommodate the cyberdeck's components, and allows the screen to tilt back to a 20-degree angle.

hsmag.cc/CyberBB

GOOD FOR GAMING

Taking your entertainment on the road

Fancy Octopus Arcades

● **Brooklyn-based Shonee Strother and his son Wolf both loved arcade games and the characters found in animations and cartoons, but the constraints of a New York apartment meant their entertainment ideas far outstripped the realities of what might be practical at home.** Shonee "saw folks making huge gaming rooms, loading them up with arcade cabinets, and realised that for some of us in smaller apartments, that wasn't a possibility." The pair set about installing RetroPie builds inside portable boom boxes and mini cabinets and customising each 'fancy' arcade with plastic dolls and other props to match each arcade's games offerings. The mini cabinets are Raspberry Pi-powered and "packed to the gills with games and easily plug in via standard HDMI for video audio output." Each design is unique, and the

mini arcades are themselves miniature art pieces that look good displayed on a bookshelf or, indeed, shown off on Instagram, which was the main reason Fancy Octopus Arcades began getting noticed. It was also a neat way of Wolf immersing himself in the games history his dad was keen for him to soak up.

hsmag.cc/OctopusArcade ⟫

Bluetooth speaker Mini PC

■ **Carter Hurd didn't really want or need a Bluetooth speaker, but having been sent one on spec, he set about making use of it anyway rather than have it sit around gathering dust.** The Divoom Ditoo Plus Retro Pixel Art Game Bluetooth Speaker's distinctive looks seemed to lend themselves to re-use as a mini PC, not least because it had several buttons and a navistick already. However, Carter dispensed with these, as well as the original speaker, in order to fit in a screen. He then built a cyberdeck computer using Raspberry Pi 3B and an old BlackBerry keyboard, which was a near-perfect fit. He covered up any rough edges and the surrounding board by 3D-printing a bezel. However, the screen was more of a challenge. "Being square was a good start, but the glass was too large to fit within the speaker housing." He ended up taking a pair of shears to the screen to trim off its edges – a rather risky endeavour – then shaped a piece of thermal plastic into a dome with a hairdryer and placed it over the LCD to give the screen a retro look. While component size versus the Divoom case was an issue all round (Carter says Raspberry Pi Zero would have been a better fit), the end result looks great and runs Raspberry Pi OS "a treat." Had he pre-planned this build, however, he'd love to have found a way of preserving the touchscreen.

hsmag.cc/MiniPC ⟫

Framework Cyberdeck

▲ **"It's weird, heavy, and looks like it came out of a portal to some alternate timeline, but I personally like it,"** says maker Ben Makes Everything of the DOOM-playing cyberdeck he built from Framework laptop parts.** His take on the concept of a cyberdeck is an individually tailored, portable computer. Framework supplies the basic hardware for the customer to build, repair, and upgrade at will, but at lower cost than other bespoke laptops. It has a modular I/O system with USB-C connections and runs Windows (and is easily capable of running many mainstream games) but also lends itself to modifications. Ben's own designs, based around the 16GB RAM laptop with Western Digital hard drive he created, can be found at **hsmag.cc/BMECyberdeck**.

hsmag.cc/FrameworkCD ⟫

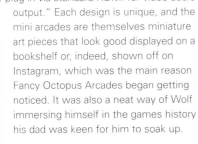

COSPLAY AND FILM PROPS

Get dressed up

Fancy dress just doesn't cut it these days. Realistic props and intricate cosplay outfits, along with studio-quality make-up, are expected in this age of constant visual scrutiny. If you're new to the world of properly dressing up in character, the Costume Wall website has some good ideas: **costumewall.com/titles**, e.g. **costumewall.com/title/adventure-time**. Perhaps you love *Stranger Things* and yearn to dress like a Demogorgon (**costumewall.com/dress-like-demogorgon**), or want to recreate a battle scene and need to make armour that passes muster, in which case Kamui Cosplay armour tutorials can help **kamuicosplay.com**. Cosplayers also rate Coscraft (**coscraft.co.uk**) for supplies.

Proton Pack Slimer

You can probably source your own *Ghostbusters* boiler suit, but hats off to the parents of Livi, who set about creating this amazing proton pack, complete with ectoplasm. Expandable foam on a piece of dowel was used to make Slimer himself, using a coat hanger to give the body form and foam-filled gloves for hands. The face, and other features, were then carved and the whole beast painted. The proton pack has three LEDs and was made from cardboard, plastic, wires, ribbon cable, stickers, a calculator, random switches and knobs, some hose, pipes, and batteries, explains mum Katrina who wrote up the family project.

hsmag.cc/slimer

Yet Another Flux Capacitor

Ambrogio is far from the first film fan to try and recreate items from 1980s classic *Back to the Future*, hence calling this project YAFC (Yet Another Flux Capacitor). A device that propels us into the near future or past seems a fitting DIY project for a keen paraglider given to flights of fancy. "The internet is full of replicas made of LED strips, but I wanted to follow a different innovative approach, reproducing the 'energy flow' effect in the flux capacitor by means of a video created with a video editor software." He used Raspberry Pi, since it has a built-in VLC player that will work with almost any inexpensive display, and supports Python. "From a maker point of view, [it] is a great language since it does not require complex environment setup," he finds. He attached a Pimoroni Automation HAT to Raspberry Pi 4 and used VSDC video editor to get the flux capacitor effects he was after. The whole thing is hidden within a painted cardboard frame and runs headless, with three large control buttons to initiate the looping flux effect (viewable as a YouTube clip). Find his instructions at **hsmag.cc/GHyafc**.

hsmag.cc/yafc

ELECTRONIC JEWELLERY

Put some spark in your accessories

Watchy

☐ **Watchy is a great-looking timepiece with a 200×200-pixel e-paper display, a glare-free 180-degree viewing angle, and ultra-low power consumption.** Powered by an ESP32-PICO-D4, it is a certified open-source hardware kit that you can customise with your own watch-face designs. Examples (and code provided) include random Tetris and mandala designs. There is also the option to design your own watch-case with the provided schematic if you don't wish to stick with Watchy's standard version. The kit includes the PCB, battery, double-sided tape, wrist strap, plastic case, and e-ink display.

watchy.sqfmi.com

NeoPixel Cosmic Turtle Necklace

◯ **There's a lot to like about this light-up pendant build which makes use of Adafruit NeoPixel Jewel, which has seven controllable LEDs, a Trinket, and your choice of either upcycled or 3D-printed and painted pendant shell.** Maker Erin St Blaine chose a pendant shell with the right sort of holes to optimally show off the illumination created by wiring up a tiny breadboard with minute backpack charger, before soldering on two lengths of wire that double as the necklace itself. Adding a micro USB connection means recharging can be done by plugging into a laptop or USB charger when needed.

hsmag.cc/LEDpend

Cyberpunk Fusion Ring

◯ **Jewellery that incorporates electronic components can make a great statement, and we've seen versions as earrings, cuff-links, watches, and necklaces.** A clever way to upcycle circuit boards and declare your techie leanings, few such accessories combine utility with fashion. However, NFC offers a little something extra (**hsmag.cc/pcbring**). Maker marketplace Etsy lists pendants that can be customised with a personalised message – essentially modern-day lockets (**hsmag.cc/nfcpend**) – while jeweller Eli Yoo meticulously crafts NFC rings that can be used to unlock a phone and make a micropayment. Eli describes his collection (**hsmag.cc/nfcring**) as Cyberpunk Fusion Rings that provide a futuristic twist and embrace the cyberpunk vibe. The precisely CNC-milled brass rings are "adorned by a circuit board, featuring an added NFC chip and tritium tube." Glowing LED vials can be set to blue or orange, while the NFC chip is programmable so the owner of this "wearable piece of the future" can select a range of functions.

Bat Girl Birkin Bag

▲ **Dennis Louie, aka Caitlinsdad on Instructables, created the Batgirl Birkin Handbag a decade ago, figuring** "What's more exclusive and has more cachet than scoring a Birkin bag? Pow!!! A Batgirl Birkin bag!" The Batgirl bag was a plasticky Birkin knock-off with a helpfully translucent coating that a "megalumens flashlight" can shine through. Customising it involved adding purple, yellow, and black fabric trim, plus a length of cable for the Taser probe and a pink toy phone handset that he wired to work with a smartphone. Caitlin's version features a bat signal, bat proximity alarm, bat taser, bat phone, and plays the *Batman* TV theme. Based on its maker's nostalgia for the 1960s TV series, the accessory is hefty enough to serve as a weapon, though the Batman the theme tune, enabled by Adafruit FLORA, is a little on the quiet side. Dennis suggests listening to its play here: **hsmag.cc/BGsound**.

VISUAL SPECTACLES

Making everything look just right

Miniature observatory

This gorgeous mini observatory is the work of Matt Hough, a retiree with a long-term interest in astronomy who now has time to put the Python and hardware design ideas he's been honing into practice, starting with experiments using Raspberry Pi Pico. He taught himself about stepper motors for use with astronomy cameras, then realised an RP2040 offered the power and control he needed and would work alongside the main Raspberry Pi 4B. "Mechanical and observatory design were something new for me. I'd done some 3D printing before, but never had to design so many parts myself. The observatory, gears, and mechanism are all homegrown. The semi-intelligent motor controller for the telescope is probably the most novel homegrown element. I needed a way for the telescope to move itself whilst the Raspberry Pi was busy taking photographs, so I gave the motors a little brain. I'm sure there are better solutions out there to some of the problems I faced – I hope to still find a few for the next version. It felt like a distraction at the time, but it's a new skill, and it has come in useful elsewhere since." He used CircuitPython for the 'very nice' Adafruit components and made use of Python libraries for the Skyfield, OpenCV, PiDNG, and Astroalign astronomy-specific features (see his GitHub: **hsmag.cc/Pilomar**). Chief among Matt's goals for the next version of the mini observatory is finding a way to have the telescope process images in real-time.

hsmag.cc/MiniObs

489MP DIY Scanner Camera

Yunus Zenichowski's prototype 489MP camera might be a bit of a misnomer: it's actually an Epson flatbed scanner capable of capturing high levels of detail. Its CCD moves across a fixed, linear area and processes an image for each line of the image. Having removed the mechanical CCD from the old scanner, Yunus attached an inexpensive projector lens to the front. He admits that this cost-saving choice "comes at a cost of sharpness." However, aside from any 3D printing costs, it means a high-resolution camera can be built from other photography-related parts. Modifying the scanner potentially results in distortion and ghosting effects due to the lengthy exposure times, but could also make for some interesting artistic effects that lend themselves to large-scale printing. An advantage is the scanner has a colour depth of 48-bit, rather than the more common 8 or 24. Yunus details his DIY idea at **hsmag.cc/489MPcam**.

hsmag.cc/megascanner

DIY Vegas Sphere

When DrZzs and GrZzs saw the Las Vegas Sphere entertainment complex, they knew they wanted to build a replica. They settled on an 8-inch Mini Megasphere (a little smaller than the 516ft-wide Vegas version), but are also "pretty far along creating a smaller version that pretty much anyone can 3D-print and build." Challenges included ensuring the pixel density was distributed uniformly around the sphere. After all, if they "just wrapped a matrix of pixels around a ball, they would have ended up with "a bunch of pixels bunched together around the top and bottom." To equally space the pixels, they ended up with 45 pixels in the top ring and 310 around the middle. The frame consists of five horizontal rings, 18 vertical ribs, and 77 rings of pixels. A commercial CNC routing company cut out all the lightweight 'gorilla-ply' parts. Weighing roughly 550lbs, the huge sphere also needed to be protected from the elements. Once painted and waterproofed, 20,000 pixels were attached to 1-inch-spaced pixel straps. Even with the help of "pixel-pushing pliers, a pixel-pushing jig, and wearing work gloves", the process took 50 hours!

hsmag.cc/VegasSphere

GET MUSICAL

Build your own instruments

Fuzz Effect

Fuzz effects can make music seem more raw and alive, rather than regular and controlled, adding reverb and delay to shape the tone to make it more unique and appropriate, in the words of maker Handy Bear. Beloved of rock guitarists, it seems fitting that the grungy image is echoed by the homemade vibes of this DIY distortion pedal build. Using a blank breadboard, some jumpers, LEDs, and capacitors, and encasing the electronics in a metal enclosure, you can make a range of effects. Sites such as **tagboardeffects.blogspot.com** advise on how to make effects reminiscent of particular songs and bands, explains Handy Bear, and a fuzz effect is fairly simple and a good starting point. Spaces for the foot switch, LEDs, potentiometer, and cable all need to be marked out and drilled before the power jack and cable socket are added (and isolated) and the enclosure is wired up as a foot pedal.

hsmag.cc/**FuzzEffect** \\\

Stepper Motor Piano

Thisinomine's gorgeous instrument works by spinning motors at different speeds. There are a few such examples around, but this one is user-playable, rather than creating random notes or playing a preset song, much like an old-time fairground organ. "I wanted to create something that looked and felt like an instrument, but secretly had an electronic skeleton," says Nomine. The motors used here have 200 steps and can be rotated a single step at a time, so they are highly accurate, and can be set to rotate a certain number of degrees and at a certain speed. This level of accuracy makes them ideal for music-making. Whenever one of the copper keys is pressed, it sets off one of the stepper motors, controlled by an Arduino. The acoustics of the piano's resulting notes are enhanced by the polished African hardwood case in which it's encased.

hsmag.cc/**MotorPiano** \\\

HackSpace
TECHNOLOGY IN YOUR HANDS

REVIEWS

HACK | MAKE | BUILD | CREATE

Cool stuff to help you make even more more cool stuff

PG 160

BEST OF BREED: ELECTRONICS PROTOTYPING BOARDS

Until you're comfortable with ordering your own custom designs direct from the factory, you'll need a prototyping board – and we've got some of the best right here

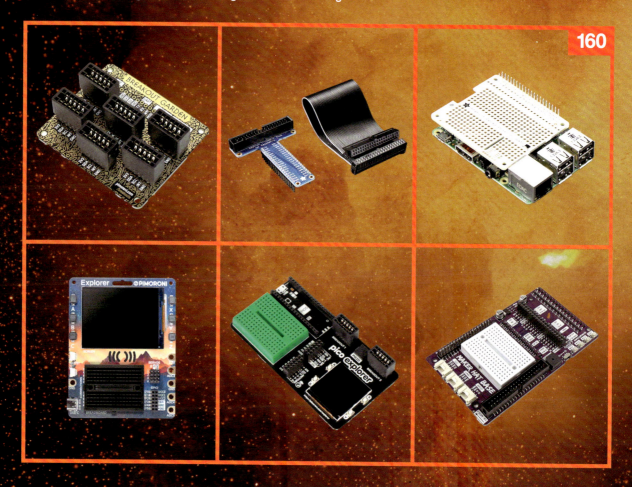

160

PG 166

PIMORONI PLASMA 2350

Illuminate masses of LEDs with this programming control board

PG 168

GEAR GUIDE

The best accessories and upgrades in the world of Raspberry Pi

ONLY THE **BEST**

Electronics prototyping boards

By **Phil King**

The GPIO (general-purpose input/output) pins on your Raspberry Pi enable you to connect electronic components and sensors as you experiment with building your own circuits on a breadboard. Since the GPIO pins aren't labelled, however, you often need to do a bit of pin counting to find the correct pin to connect to while referring to a pinout diagram. That's where breakout boards can come in useful: they make it easier to connect electronics and focus on building your projects.

Ever since Raspberry Pi was launched in 2012, third-party manufacturers have come up with a variety of breakout boards to help digital makers. They range from the very basic – 'cobblers' that use a ribbon cable to connect the GPIO pins directly to the rows on a breadboard – to more advanced boards with bonus features such as status LEDs, extra connections, input devices, and mini LCD screens.

We take a look at some of the best – including one for Raspberry Pi Pico, along with a standalone RP2350-based board – for exploring electronics prototyping.

The Pi T-Cobbler is supplied with a ribbon cable

Adafruit Pi T-Cobbler Plus

Adafruit | £8 / $8 | adafruit.com

The original T-Cobbler was launched by Adafruit in 2012, the same year as the very first Raspberry Pi! The expanded 'Plus' version works with all 40-pin models (i.e. all modern Raspberry Pi computers), but the concept is exactly the same: it's a handy way of breaking out the GPIO header pins onto a breadboard for the prototyping of electronic circuits.

The T shape enables the supplied ribbon cable to be connected between Raspberry Pi's header and the T-Cobbler without obstructing access to a breadboard. The bottom part of the T features 40 downward-facing pins to fit into the breadboard's holes – either side of its central 'ravine'. So, by inserting the end of a jumper wire to a hole in the same row, in effect you're connecting it to the respective GPIO pin. Not only is this handy, but – unlike on Raspberry Pi – all the pins are labelled, so it's a lot easier to find the one you want.

Verdict

An inexpensive way of making GPIO connections a lot easier.

Maker HAT Base

Cytron | £11 / $11 | cytron.io

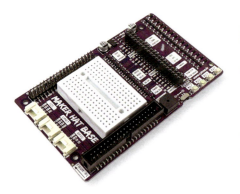

◄ There's a lot packed into this very useful board

Owners of a Raspberry Pi 400, 500, or 500+ needn't miss out when it comes to breakout boards. While you can connect any HAT to its rear-mounted GPIO header, the top of the HAT will be facing the rear – not very user-friendly when you're trying to read the labels for connecting things up to a breakout board.

Designed specially for Raspberry Pi 400, 500 and 500+, the Maker HAT Base makes things easier. A supplied short ribbon cable connects it to Raspberry Pi's GPIO header, so the HAT itself can lie flat on your desk or table. You might also find it useful for extending the GPIO header of a standard Raspberry Pi model enclosed in a case.

It's equipped with a good deal of functionality, too, including onboard push-buttons, a buzzer, and a mini breadboard, along with a female breakout header with helpful pin labels and status LEDs. There's also a fully labelled male 40-pin GPIO header.

Breakout Garden

Pimoroni | £13 / $15 | pimoroni.com

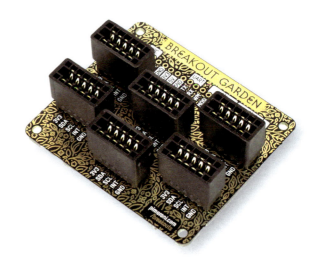

▲ There are six slots to insert tiny breakout boards

Adding sensors and other devices to a Raspberry Pi project via a breadboard is all very well, but sometimes you just want to get on with making and coding. In this case, the Breakout Garden makes it a lot easier. The standard HAT version comes with six sockets to insert any of the compatible tiny breakout boards available. There's an ever expanding array to choose from (**rpimag.co/bgcollection**), including all manner of sensors, plus LED matrices, tiny LCD screens, and even thermal cameras.

It really is plug-and-play, with each breakout's fixed I2C address recognised by the accompanying software, so it'll detect them correctly even if you move them around. And, since they use a shared I2C bus, this leaves plenty of spare GPIO pins, the most useful of which are broken out by a 20-pin strip of holes. Other Breakout Garden models are also available: the three-slot Mini and Raspberry Pi 400/500/500+ versions, plus a six-slot Base for Pico.

Adafruit Perma-Proto HAT

Adafruit | From £5 / $5 | adafruit.com

Experimenting with circuits on solderless breadboards is fun, but there may come a time when you want to create something more permanent for a project. This is where protoboard comes in useful, enabling you to solder components and wires to create more reliable and durable circuits.

The Perma-Proto HAT kit takes it one step further by mounting a piece of protoboard on top of Raspberry Pi itself. Compatible with any 40-pin Raspberry Pi computer, the standard version can even be stacked with another regular HAT. Alternatively, choose the version with a built-in EEPROM that you can program to make a setup using Raspberry Pi HAT specs – in effect creating a custom HAT.

Either way, the protoboard features traces on the rear to connect the rows of holes in the same way as a standard breadboard, with a familiar layout. There's also a handy breakout row for the GPIO pins, along with power and ground rails.

▲ Make your own custom HAT with this kit

Verdict

A great-value way to build more permanent circuits.

Pimoroni Explorer

Pimoroni | From £34 / $38 | pimoroni.com

◄ An RP2350-based electronics prototyping board that's very expandable

Verdict

A great way to get started with electronics prototyping.

Not to be confused with the classic Explorer HAT Pro (still a great option for Raspberry Pi computers), this standalone Explorer is based around the RP2350 microcontroller chip – as used in Raspberry Pi Pico 2 / 2W, so it's programmable with MicroPython and C/C++.

Billed as an 'electronics project playground', it features a mini breadboard next to a female header breaking out a selection of RP2350's GPIO pins, including three analogue inputs, along with 3.3V and a couple of GND pins. So it's easy to add electronic components to circuits on the breadboard and connect them to the pins. There's also a header to connect up to four servos, along with a couple of Qw/ST ports and six crocodile-clip terminals.

The highlight is a 320×240 2.8-inch LCD screen, ideal for displaying data and graphics. There's even a tiny piezo speaker for basic audio. The Explorer Starter Kit version contains a range of electronics components to get started, including a Qw/ST multi-sensor stick and a couple of motors with wheels.

Pico Explorer Base

Pimoroni | £25 / $28 | pimoroni.com

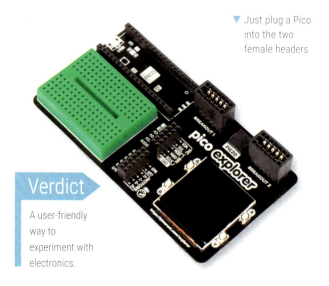

▼ Just plug a Pico into the two female headers

Verdict

A user-friendly way to experiment with electronics.

Raspberry Pi Pico users aren't left out when it comes to breakout boards. The Pico Explorer Base is one of the best, featuring a mini green breadboard with 170 points, although you could always use a larger separate one if it proves a little cramped for your projects. A mini LCD screen is useful for displaying data and is equipped with four tactile buttons.

A selection of Raspberry Pi Pico's pins are broken out via two 12-pin female headers. Clearly labelled, these include I2C, SPI, ADC, and seven standard GPIO pins, along with two GND connections and one 3V3 power. An Audio pin is connected to the onboard piezo speaker.

As a bonus, the four remaining breakout pins are allocated to motor connections. Making use of a DRV8833 dual H-bridge motor driver chip, these can deliver 1.5A RMS current output to control two DC motors (or other power-hungry devices such as NeoPixels).

ELECTRONICS KIT

Various | From £6 / $7 | various

While you can buy all manner of electronics components separately, a kit is a more cost-effective option for beginners. There are countless kits available, ranging from the inexpensive entry-level CamJam EduKit #1 to others comprising hundreds of components, such as the SunFounder Super Starter Kit V3.0 for Raspberry Pi (pictured). They typically come with a series of lessons to help you to get started.

▼ This SunFounder kit offers a huge array of components

Plasma **2350**

▶ Pimoroni ▶ **magpi.cc/plasma2350** ▶ From £12 / $13

An LED strip controller with even
more processing power. By **Phil King**

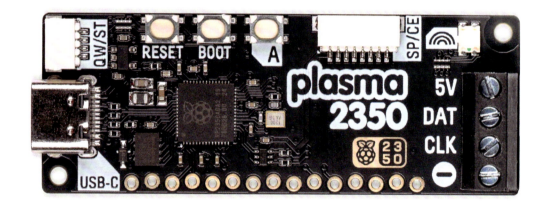

SPECS

FEATURES:
RP2350A
processor; Boot,
Reset, and user
buttons; on-
board RGB LED

**LED
COMPATIBILITY:**
5V WS2812/
NeoPixel,
APA102/DotStar

CONNECTIONS:
4 × screw
terminals, USB-C
port for power/
programming,
Qwiic/STEMMA
QT, SP/CE,
unpopulated 15-
pin GPIO header

Verdict

An easy way to
control NeoPixel/
DotStar LED strings
with programmable
effects, with extra
processing power
if you need it.

9/10

▲ Only slightly longer than a Raspberry Pi Pico 2, the
Plasma 2350 makes use of the same microcontroller
chip for fast processing and flicker-free lighting

▶ The Starter Kit includes a 10m string of frosted RGB LED stars
that showcase the Plasma 2350's capabilities with some eye-
catching effects

Featuring the same RP2350 microcontroller
chip as Raspberry Pi Pico 2, the Plasma
2350 can illuminate strings of 5V WS2812/
NeoPixel or APA102/DotStar individually
addressable RGB LEDs with some eye-catching
lighting effects.

Based on a Dual Arm Cortex M33 running at up
to 150MHz with 520kB SRAM, the processor is a
major upgrade on the RP2040 used in previous
Pimoroni products such as the Plasma 2040 and
Plasma Stick 2040 W. While lighting LEDs isn't
the most demanding of tasks, the extra processing
power may come in useful for more advanced
projects linked to breakout boards. There's
double the amount of QSPI flash storage (4MB)
to play with, too.

You can buy the Plasma 2350 board on its own
or in a Starter Kit (£34.50 / $38) with a USB-A
to USB-C cable and 10m string of 66 individually
addressable frosted LED stars. These look superb
when lit up, and are a great way of showcasing
the capabilities of the Plasma 2350. We also tried
out a long 300-LED strip and there was easily
enough current (up to 3A) from the USB-C power
connection to light them all.

Wired for light

Connecting your LED string or strip to the board
is simple. As on the Plasma 2040, there are four
screw terminals on one end: for 5V power, data,
clock, and ground. While WS2812/NeoPixel LED
strips have only three wires, omitting the clock

connection, the latter is needed for DotStar LEDs. A little care needs to be taken to make sure each wire is in the correct terminal and that they're screwed securely.

Despite coming in a slimmer 'gum stick' form factor than the Plasma RP2040, the board manages to cram in many useful features. There's an unpopulated header down one long edge to break out selected GPIO pins, offering access to UART and

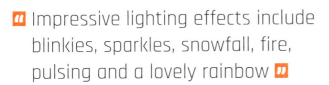

> ❝ Impressive lighting effects include blinkies, sparkles, snowfall, fire, pulsing and a lovely rainbow ❞

I2C interfaces, along with analogue inputs, PWM outputs, and PIO state machines. There's also a Qwiic/STEMMA QT connector for breakout boards, as well as Pimoroni's new proprietary SP/CE (Serial Peripheral / Connector Evolution) port – an eight-pin JST-PH connector including four pins for SPI.

Buttons for everything

One slight downside of the SP/CE's inclusion is that there's no room for a B user button next to the A one (as on the Plasma 2040), although the Boot button doubles as a user input. There's also a handy Reset button, to save repeatedly disconnecting and connecting the USB-C power. An on-board RGB LED is a nice touch, too.

Before you can start programming some light patterns, you'll need to install MicroPython. In its RP2350 GitHub repo (**magpi.cc/pimoroni2350gh**), Pimoroni provides a custom UF2 file for the Plasma 2350. To install it, connect the board to your computer via USB while holding the Boot button, to mount it as a drive, then drag the file over to it. Alternatively, if you prefer CircuitPython, with which you can utilise Adafruit's excellent LED Animation library, there's

also a UF2 image for that: **magpi.cc/plasma2350cp**.

Using MicroPython in the Thonny IDE, we tried out some code examples from Pimoroni's GitHub repo, altering the constant for the number of LEDs to match our string. While there are only a few examples there – including a nice 'rainbow' colour-cycling effect – we found that most of the ones for the Plasma 2040 and Plasma Stick 2040 W still work. Impressive lighting effects include alternating/random blinkies, sparkles, snowfall, fire, pulsing and a lovely rainbow sweeping across the string of LEDs.

Since the Plasma 2350 lacks on-board Wi-Fi, you can't get your LEDs to react to data from the network, but you could always connect a breakout input such as a temperature sensor ◾

▲ The rear of the Plasma 2350 board, showing helpful labels for the LED strip screw terminals and unpopulated GPIO header

▼ The Plasma 2350 packs a lot into a small footprint, including buttons and an on-board RGB LED

Gear Guide

Discover a treasure trove of Raspberry Pi devices and accessories

Rosie Hattersley

Whether, like our very own Features Ed Rob, you regard Christmas as the most wonderful time of the year, or it's simply a period when you get a well-earned break and can finally focus on favourite hobbies, there's lots to love about the tail end of the year. Festivities over, you can clear the decks and spend some quality time with shiny new tech treats, perhaps including some Raspberry Pi gems that will make your life run that much more smoothly and your leisure hours much more fun. Here's our guide to what you should put into your shopping basket for next year.

Buy a Raspberry Pi

Raspberry Pi is really spoiling you with a choice of computing options to suit your pockets as well as your projects

Raspberry Pi 500+

→ rpimag.co/500plus
→ From $200

There's simply no beating the amount of computer that Raspberry Pi 500+ gets you for your money. It's a Raspberry Pi 5, with 16GB RAM, a 256GB solid state drive, and a gloriously clicky backlit RGB mechanical keyboard, in one delicious package. The 2.4GHz Raspberry Pi 5 inside has an exposed 40-pin GPIO header, so plenty of accessories can be attached to the back of the keyboard too.

Raspberry Pi Zero 2 W

→ magpi.cc/zero2w
→ £17 / $15

Make your home smarter with the Pi Zero 2 W: it is ideal for IoT projects such as motion detectors and security cameras. The tiny Raspberry Pi is both wireless and Bluetooth enabled, so can be used around your home and in small, power efficient projects. With 512MV onboard memory and a 64bit Arm processor, it's a brilliant choice for gaming, drones and tiny robot builds too, see these projects: **magpi.cc/zeroblogs**.

Raspberry Pi 5

→ magpi.cc/raspberrypi5
→ £48/$50 (2GB); £57/$66 (4GB); £77/$82 (8GB); £115/$132 16GB

The flagship Raspberry Pi model sports a 2.4MHz quad-core processor, PCI Express and up to 16GB of RAM for the smoothest, fastest computing experience to date. As well as being a great platform for the AI applications, Raspberry Pi 5 is more than twice as powerful as Raspberry Pi 4. It sports USB 3, includes nippy PCI Express with the bandwidth to make proper use of SSD, a real-time clock, gigabit Ethernet, and dual 4K digital display connections. It even has a power button.

Raspberry Pi Pico 2

→ magpi.cc/pico2
→ From £4.80 / $4.99

This souped-up version of our powerful microcontroller, Pico 2 features PIO (programmable input/output– see **magpi.cc/whatispio**), optionally comes with wireless connectivity and is ideal for Internet of Things projects to make your home super-smart. The RP2350 chip is faster, has a dual Cortex Arm or RISC-V cores and twice as much onboard RAM as the original Pico.

Accessory guide

We've been rolling out lots of lovely Raspberry Pi add-ons over the past few months. What's not to like?

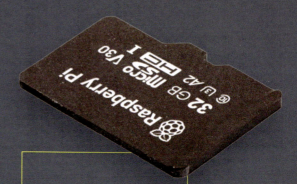

A2 SD Card

→ magpi.cc/sdcards
→ From £10 / $13

Our very own branded Class A2 SD cards have been specifically with support for DDR50 and SDR104 bus speeds and command queueing (CQ). You'll need to update to the latest version of Raspberry Pi OS to take full advantage of the extra performance a Class A2 SD card offers. For the technical details see **magpi.cc/sdnbumper**.

Raspberry Pi SSD Kit

→ magpi.cc/ssdkit
→ From £37 / $40

While SD cards have traditionally been used for Raspberry Pi OS, there has long been a clamour for faster, more capacious solid state drives. The new Raspberry Pi SSD Kit comes in 256GB or 512GB versions and is packaged with an M.2 HAT+ for lightning-fast disk access and Raspberry Pi boot-up times.

Raspberry Pi Bumper

→ magpi.cc/bumper
→ £3 / $3

Protect your precious Raspberry Pi with a cute silicon-rubber surround that ensures knocks don't push it off course. Available in translucent white or solid black, the Raspberry Pi Bumper is a great alternative to a case, ensuring plenty of natural airflow to keep things cool while showing off the computer at the heart of your DIY projects.

M2+ HAT

→ magpi.cc/m2hatplus

→ £12 / $12

For the fastest data transfers, an M2 SSD that uses the NVMe protocol is just the ticket. Raspberry Pi's official M2+ HAT sits atop Raspberry Pi 5, adding M2's speedy 500MBps transfers to and from NMVe drives as well to the board's existing PCIe 2.0 interface. If AI projects are your thing, you'll really notice the performance difference when using accelerators.

Touch Display 2

→ magpi.cc/touchdisplay2

→ £58 / $75

This seven-inch capacitive touchscreen boasts an 800×480-pixel IPS display with support for five-finger touch and an onscreen keyboard. The screen connects to your Raspberry Pi via an adapter and can be powered by a ribbon cable with DSI connector and a GPIO with no need for an external power pack, so it can be used as a standalone display for entertainment or IoT duties.

PicoZX Handheld

→ magpi.cc/picozxhandheld

A tiny gaming device inspired by Sinclair's ZX Spectrum could be the perfect project for those of a certain age who grew up learning computing on the home computing visionary's original. The PicoZX Handheld consists of several custom circuit boards soldered on to a Pico, plus a retro faceplate that conceals the charger, battery and connectors, USB ports and SD card slot.

ePiPod

→ magpi.cc/epipod

Del Hatch's iPod-style music player encases a Raspberry Pi Zero 2 W in a homebrew case (find the STL files here: **magpi.cc/epipodstl**) with a Waveshare ePaper touchscreen for the display and controls. Once assembled, you will have a portable player that serves up albums from an SD card, doesn't need to be connected to the internet to play them, and offers more than five hours of entertainment before needing a recharge.

Projects

Air Hockey table/ games emulator

→ magpi.cc/minihockey

Maker Chris Downing and his @BitBuilt gaming site co-owner CreshBash spent more than two years designing, 3D-printing and perfecting this 1:5 scale air hockey table. It offers either regular air hockey (first to seven points) or Doom mode for a fiery table lightshow complete with scream effects. Since it uses Retropie there's also a games emulator mode. The whole crazy caboodle is controlled by Raspberry Pi 4.

Pretty Tide Clock

→ magpi.cc/prettytideclock

This project was inspired by maker Levi wanting to know when to descend his cliff-top home and expect to swim. Even if you have no pressing need to know when and where it will be high or low tide, this gorgeous Raspberry Pi 3B+ project pulls tidal data from the US National Oceanic and Atmospheric Administration API. The 3B+ converts them into PWM values, used for the analogue gauges. The tide clock indicates a rising and receding tides via red and green LEDs either side of its lovely wooden frame.

Argon Neo 5 case

→ magpi.cc/neo5
→ £18 / $23

Protect your Raspberry Pi from knocks and accidental spills with this hardy case – the latest in Argon's well-regarded range of cooling accessories for Raspberry Pi is well geared-up for school or industrial use. The aluminium enclosure offers effective CPU and circuitry cooling helped by an onboard, Pi-controlled 30mm PWM fan which is largely silent. Raspberry Pi's SD card slot can be left accessible when the enclosure is in place, or optionally covered with a screw-on plate to stop anyone messing with your setup.

> "The aluminium case offers excellent cooling qualities"

CrowView Note

→ magpi.cc/crowview
→ £128 / $169

Slide your Raspberry Pi into the side of the CrowView and you've got your very own sliver of portable computing convenience. This laptop dock has a socket that is designed especially for Raspberry Pi 5, but can also be used with earlier versions of our favourite computer. Once connected, Raspberry Pi OS runs on the CrowView without further prompting. The dock has a 5000mAh battery pack which powers both dock and Raspberry Pi for up to two hours

Pironman 5 case

→ magpi.cc/pironman5
→ £70 / $91

With its transparent sides and internal spot lights, plus detailed LED status info, the Pironman 5 reminds us of PC computing kit beloved by modders and overclockers. The aluminium case offers excellent cooling qualities for optimal performance. There are USB 2.0 and USB 3.0 ports, gigabit Ethernet and two HDMI ports and support for both M2 SSD hard drives and has a Raspberry Pi 5 NVMe PCIe peripheral board. Geek out!

Tiny Circuits Thumby Color gaming device

→ magpi.cc/thumbycolor
→ £38 / $45

While the initial appeal of Thumby Color probably lies in being able to play all sorts of arcade-derived and board games on its teeny 128×128-pixel 16-bit colour screen, the RP2350 microprocessor-powered Thumby soon lures you in to the world of coding your own games. Reviewing it, *MagPi* editor Lucy remarked that "The real deal is the ability to investigate the API and create games yourself by following the tutorials [to which end] Thumby has an online Code Editor and a starter guide. Thumbs up all round!

Cool kits & electronics

Tinkering with lights, sensors and diodes
is a great way of exploring electronics

Raspberry Pi Pico Advanced Kit

→ **magpi.cc/advpicokit**
→ **£29 / $38**

Electronics kits are something
of a classic festive gift, and
we can certainly see ourselves getting stuck
in to this absolute trove of jumpers, diodes,
LEDs, breadboards, sensors, inputs, outputs,
robot components and bleepy things. An ideal
companion set for Pico owners of any
age, it also comes with instructions
for several MicroPython Pico projects.

Raspberry Shake

→ **raspberryshake.org**
→ **From £267 / $349**

Citizen scientists will love this incredible device that can measure ground
tremors near and far. From volcanoes and earthquakes another continent
away to heavy traffic and even the footfall of large crowds, all sorts of seismic
events are recorded on Raspberry Shake, and can duly be seen by fellow
shake-watchers around the globe. Depending on the version you choose,
Raspberry Shake records tremors across one, two or four axes, (vertically
and laterally) including underground, under the sea and even overhead.

Raspberry Pi AI Camera

→ **magpi.cc/aicamera**
→ **£63 / $70**

Perhaps the most prevalent tech phrase of recent
years (closely followed by 'blockchain'), AI
promises to radically alter how we interact, glean
information, process data and create and consume
images. We couldn't resist designing our own AI
camera based around Sony's IMX500 Intelligent
Vision Sensor. With its on module AI processor,
the Raspberry Pi AI helps you create impressive
vision AI applications and neural network models.
There's a useful Getting Started guide at **magpi.cc/
aicamdocs**.

The 12.3Mp IMX500 Raspberry Pi AI Camera
includes everything you need for Raspberry Pi
photography and works with the AITRIOS platform
(**magpi.cc/aitrios**) to create your own classifier
detectors, train and label models, and produce
brilliant machine-learning projects.

Amazing maker tools

A fully loaded box of tricks will see you through many making adventures

Solder Scroll

→ magpi.cc/solderscroll
→ Free to 3D print

Hackspace's Andrew Gregory immediately saw the use of Solder Scroll, describing it in his review as a "brilliant device [that] makes soldering more ergonomic by allowing you to dispense any diameter of solder out of an object you hold like a pen". It effectively provides an extra hand to accurately hold in place that tiny object you're attempting to solder. You just need to 3D-print the handy tool using designer Victor's STL files: **magpi.cc/scrollstl**.

Hozo NeoRuler GO

→ magpi.cc/neorulergo
→ £46

This digital rolling ruler is accurate to less than 1mm and provides measurements in inches, centimetres, feet, yards and miles. However, it will probably be of most use to *The MagPi* and *Hackspace* readers for tackling curves and awkward shapes when planning 3D and electronics projects, schematics and drafting plans. It has 93 built-in scales. Measurements are automatically saved to the linked Meazor app.

Pimoroni Explorer

→ magpi.cc/pimexplorer
→ From £34 / $44

Pimoroni's 'electronic adventure playground for physical computing' is available both as a kit with or without a RP2350 (Pico 2) included, the Explorer goody bag includes sensors to measure moisture, temperature, light and movement, two 60mm wheels, rotation servos and a potentiometer. There's also a breadboard, 2.8in LCD display, speaker, STEMMA connectors and a generous number of analogue and digital connectors.

> "Measurements are automatically saved to the linked Meazor app"

xTool S1 Laser Cutter

→ magpi.cc/xtools1
→ £1799 / $2334

Laser cutting is a very compelling alternative to 3D printing, involving far less plastic, (although it works on this as well as wood) but also requiring proper precautions. The fully enclosed xTool S1 keeps its potentially dangerous 2W infrared and 20W and 40W blue lasers away from the user, while offering admirable engraving and cutting features. Powerful enough to use with 10mm wood or plywood, this is a great hobbyist machine.

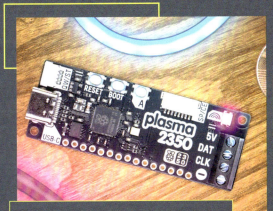

Plasma 2350 LED Light Controller

→ magpi.cc/plasma2350
→ £12 / $13

Some Raspberry Pi projects offer real coding challenges, but many smart lights and sensor-based ones can be done with a straightforward electronics setup and a suitable microcontroller. The Plasma 2350 is based around the same chip as Raspberry Pi Pico 2 and can illuminate strings of NeoPixel lights or separate RGB LEDs, making for a great fun project that's ideal for those new to coding.

CrowPi Educational Kit

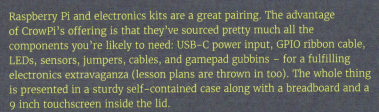

→ magpi.cc/crowpi
→ £165 / $208

Raspberry Pi and electronics kits are a great pairing. The advantage of CrowPi's offering is that they've sourced pretty much all the components you're likely to need: USB-C power input, GPIO ribbon cable, LEDs, sensors, jumpers, cables, and gamepad gubbins – for a fulfilling electronics extravaganza (lesson plans are thrown in too). The whole thing is presented in a sturdy self-contained case along with a breadboard and a 9 inch touchscreen inside the lid.

Bullfrog Synthesizer

→ magpi.cc/bullfrog
→ £450 / $541

Home computers and electronic music have plenty of shared history, which Bullfrog's subtractive and analogue synth continues. An RP2040 microcontroller takes care of MIDI implementation and is used in a sampler/Loloper voice card. Both digital instrument and educational tool with a 70-page instruction manual, it's a wonderful machine on which to experiment with pitch, timbre and amplitude. Co-designer and DJ/music producer Richie Hawtin proclaims the goal "to both nurture a passion for electronically produced sounds and promote fun".

PiDP-11 replica kit

→ magpi.cc/pidp11
→ $297

Fans of early computers and electronic music may well be familiar with DEC's PDP machines. We covered a Raspberry Pi-enabled rebuild project last issue (**magpi.cc/147**). This gorgeous replica kit is the next best thing, miniaturising the 1975 model to 6:10 with Raspberry Pi 2, 3, 4 or 5 (optionally) running the Blinkenbone SimH emulator.

ArmPi FPV AI Vision

→ magpi.cc/armpifpv
→ £200 / $300

If you want to give AI a try, you could do far worse than experiment with this programmable robotic arm which you can variously challenge to pick up specific coloured blocks, manoeuvre the arm and transport them from place to place via an app. The arm's six degrees of lateral movement include joints and a robotic 'wrist', while smarts include reporting on temperature, voltage and position.

"It supports TensorFlow and Pytorch"

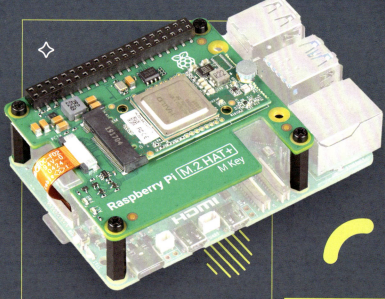

Pimoroni Inventor 2040 HAT Mini

→ magpi.cc/inventorhatmini
→ £24 /$25

Two motor ports and four three-pin servo headers make this HAT ideal for robot projects, for which Pimoroni provides plenty of relevant guidance in this package. However, the inputs for sensors, ADC-compatible GPIO connections, audio and UART ports make this a fairly flexible package for all wannabe inventors.

Raspberry Pi AI HAT+

→ magpi.cc/aihat
→ From £65 / $70

Machine learning and Raspberry Pi work flawlessly together in this integrated AI accelerator. The Raspberry Pi AI HAT+ features a built-in neural network accelerator, turning your Raspberry Pi 5 into a high-performance, accessible, and power-efficient AI machine. Available with 13 or 26 TOPS performance, it's suited to everything from entry-level applications to complex neural processing.

CM4 XGO Lite Robot Dog Kit

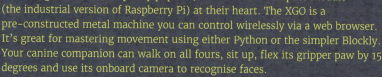

→ magpi.cc/cm4xgo
→ £429

Robot dogs have become pretty popular, but few have Compute Module (the industrial version of Raspberry Pi) at their heart. The XGO is a pre-constructed metal machine you can control wirelessly via a web browser. It's great for mastering movement using either Python or the simpler Blockly. Your canine companion can walk on all fours, sit up, flex its gripper paw by 15 degrees and use its onboard camera to recognise faces.

Mini Desktop PC

→ magpi.cc/minipc
→ 3D-printed file

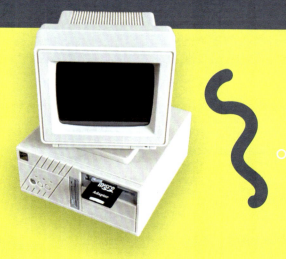

A tiny version of an MS-DOS 286 PC, this charming computer can be assembled by 3D-printing a replica case (**magpi.cc/retropccase**) and adding a Raspberry Pi plus printed circuit board, 3.5in LDC, power supply and cables. A lovely touch with this nostalgia-inducing build is that its DOSbox emulator software runs from an SD card you pop in what, on the original computer, would have been its floppy disk drive.

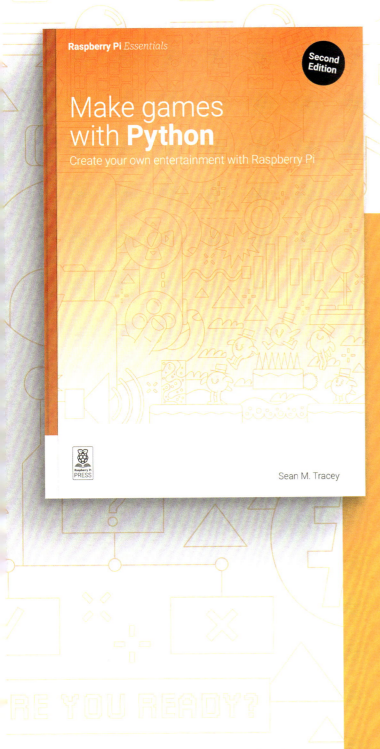

The official
Raspberry Pi
Handbook
2026

200
INSPIRING
PROJECTS!

Robot explorers
CoolCoral
Andotrope
Poetry cam
Cave Mapping
Pixie Clock
Private cloud server

FEATURING
Raspberry Pi 500+

FROM THE MAKERS OF RASPBERRY PI OFFICIAL MAGAZINE